T0306033

A Brain-Friendly Life

Modern life is brain-unfriendly: We are flooded with information and excessive cognitive demands, when we are often already depleted from chronic stress, sleep deprivation, and health issues. Many of us experience frequent "glitches" or memory lapses, despite tests showing there is nothing wrong with our brains. This book provides concrete strategies, derived from neuropsychological science and clinical practice, to help people improve how they function in daily life.

Menchola draws on her experience as a clinical neuropsychologist who has worked with a widely diverse group of patients, to translate the findings from highly controlled research into concrete strategies that people can implement in their messy worlds to make their days more brain-friendly. The book also provides advice on how to address those factors that drain our brain resources, and gives guidance on when and how to seek a neuropsychological evaluation.

It is valuable reading for anyone experiencing frustrating cognitive problems that are not due to brain disease. It is also essential for neuropsychologists, psychologists, and physicians in primary care, psychiatry, and neurology, who need a resource to offer to patients to help their healthy brains function better.

Marisa Menchola, Ph.D is a board-certified clinical neuropsychologist and Associate Professor in the Clinical Psychology Program at Midwestern University in Glendale, Arizona (United States). She also has a forensic private practice and consulting office in Tucson, Arizona.

A Brain-Friendly Life

How to Manage Cognitive Overload and Reduce Glitching

Marisa Menchola

Routledge
Taylor & Francis Group

LONDON AND NEW YORK

Designed cover image: getty images @ Dimitri Otis

First published 2025
by Routledge
4 Park Square, Milton Park, Abingdon, Oxon OX14 4RN

and by Routledge
605 Third Avenue, New York, NY 10158

Routledge is an imprint of the Taylor & Francis Group, an informa business

© 2025 Marisa Menchola

British Library Cataloguing in Publication Data
A catalogue record for this book is available from the British Library

Library of Congress Cataloging-in-Publication Data
A catalog record has been requested for this book

ISBN: 9781032529424 (hbk)
ISBN: 9781032529400 (pbk)
ISBN: 9781003409311 (ebk)

DOI: 10.4324/9781003409311

Typeset in Optima
by Taylor & Francis Books

Contents

Figures

Acknowledgments

The ideas in this book were shaped over years of clinical work at the University of Arizona Medical Center, later Banner – University Medical Center Tucson, with thousands of patients and families in the departments of Family and Community Medicine, Psychiatry, and Neurology. I put together the proposal with support from the American Academy of Clinical Neuropsychology's Author Accelerator Program. I am grateful to Farzin Irani, Ph. D., and Anthony Stringer, Ph.D., program co-chairs, for the valuable resources they put together, and to my mentor, Vonetta Dotson, Ph.D., for providing the magic touch to move the proposal forward. It was through that program that I was connected with Lucy Kennedy, Senior Publisher at Routledge, to whom I'm indebted for her graciousness and guidance through the process.

This book would not have been written without the support of many others: David Labiner, M.D., my Chair at the Department of Neurology at Banner University Medical Center, gave me time off to rest and regroup and it was during that time that the proposal became a reality, away from the unrelenting pace of full-time clinical work. Thank you for stepping in with kind and generous support during key times in my career. After moving to Midwestern University in Glendale, it was another leave that allowed me to finish this work: Thank you to Adam Fried, Ph.D., Director of the Clinical Psychology Program, and Jared Chamberlain, Ph.D., Dean of the College of Health Sciences. This book would not have been completed without their support and that of my colleagues who took on extra labor during my absence. I'm very grateful to be part of this team.

Thank you to my sisters, in order of sisterhood seniority: Mariana Menchola, Amanda Fanniff, Ph.D., Jill Lany, Ph.D., Catherine Shisslak, Ph.D., and Kristen Moore Bennett. I could not have done the last 20 years without you. Thank you to Al Kaszniak, Ph.D., to whom I owe this career I love. To Steve Rapcsak, M.D., thank you for the endless conversations about everything from semantic memory to the mid-fusiform. Your passion for this work, boundless intellectual curiosity, and reliably brilliant insights never fail to rekindle my excitement when cognitive overload threatens to drain the joy. From Cottonwood to Balaton, it's always the best time. Above everything and all, thank you to my dear Nico. It is not easy being raised by a parent who is often depleted by their own brain-unfriendly life, and you have been invariably kind and understanding. Thank you for the joy, the laughs, and the hours of passionate, caffeinated debates. I wish you a brain-friendly, soul-friendly life that you never need reprieve from.

* * *

As I was finishing this book, my father passed away, fifteen years after being diagnosed with Alzheimer's disease. This book is dedicated to every person and family that has been robbed of the chance to write a final chapter by a disease that extinguishes the very organ that holds the memories and emotions that make us who we are and bind us together.

Introduction

Depleted Brains in a Demanding World

It goes something like this:

> "Okay, let's get started. What's your understanding of why your doctor wanted you to have this neuropsychological evaluation?"
>
> "Actually, I was the one who asked for a referral for memory testing. I just feel there's something really wrong with my brain."
>
> "Tell me what you've been noticing."
>
> "Gosh, just...everything. I keep forgetting things I have to do—just yesterday I paid yet another late fee on another bill. I can't keep up at work, and twice now I have submitted reports with glaring errors on them. My boss caught them, and it was so embarrassing. Last month, I was picking my sister up from the airport and I completely missed my turn. I just kept driving, and by the time I realized I was going the wrong way, I had to do this big loop back. I feel like I can't think on the spot: I can't think of the right word, or I'll be trying to write in the tip after a meal and I just sit there, looking at the numbers, stumped. And the one that really got to me—one afternoon, the doorbell rings, and it's my neighbor, coming to pick me up because we were going to go visit our other friend at the hospital. I'd completely forgotten."
>
> "This all sounds very frustrating. And these are things that did not use to happen before?"
>
> "Not at all. I *never* missed a deadline. I was so good with numbers. I never had to write anything down. And now I just feel so..." She tears up. "...dumb. I just can't think."
>
> Her worry is obvious, and I try to sound reassuring. "That's why we're here, to figure out what is going on. Let's see... It says here that you already had a brain MRI?"
>
> "Yes, it was normal. They did blood tests, too. Everything has been normal. They keep saying everything is *fine*."
>
> "Well, that's good news. Can you think of anything that was going on right before you started having these issues? Like, an illness, or a big change in your life, for example?"

DOI: 10.4324/9781003409311-1

"It's hard to say; there's been so much going on. Right after we moved here, our daughter separated from her husband. So she and the kids moved in with us and now she's dealing with the court and the custody nightmare. My husband lost his job at the beginning of the pandemic, and his new job is over an hour away, so he can't really help. Then my mom had a stroke, and she was supposed to go to rehab, but her insurance denied it. She was doing fine at home, and I was driving out there every couple of weeks, three hours each way, to give my dad a break. But now my dad fell and broke his hip. So I've been spending a lot of time on the phone with the home health agency and my sisters, trying to come up with a plan for his surgery."

"Wow. You've been under a *lot* of stress."

She rubs her nose. "Yes."

"How has your mood been?"

"Fine." She starts crying, and I move the tissue box we keep on the desk closer to her. "It's not like I'm depressed, or anything. I just… I can't think. I'm afraid that something is seriously wrong with my brain. My dad's mom had dementia. Alzheimer's, I think. My mom had mini-strokes before the big one. I'm terrified of something like that."

* * *

Patients like this are referred to me for a neuropsychological evaluation for *cognitive complaints*. They are experiencing significant difficulties in daily life not because of physical problems, but because of problems with their memory, attention, language, thinking, etc. They forget to pick up their children from practice. They show up to a Thanksgiving dinner to realize with horror that they forgot they were supposed to bring the dessert. They can't think of a common word or they mix words up. They make repeated mistakes at work or with simple chores at home. They miss their mother's birthday. They miss their turn on their way home. They skip a critical ingredient when making a recipe they have made for decades. The front desk clerk at the pediatrician's office asks them for their child's date of birth and they stand there, frozen, unable to think of it.

These patients share their concerns with their physicians, who might refer them for a series of tests: blood tests to rule out medical problems, a neurological consult and brain imaging to rule out neurological causes, and a neuropsychological evaluation to see if tests reveal evidence of cognitive impairments. Many of them undergo all these evaluations and are ultimately told that all of their results are normal. Their doctors, me included, send them on their way with reassurances that they are perfectly "fine" and with encouragement to focus on "lifestyle factors": Get enough sleep. Eat a healthy diet. Exercise.

And yet this fails to reassure the patients, who remain convinced that there is something wrong. My physician colleagues who refer patients to my clinic—most commonly neurologists, primary care physicians, and psychiatrists—sometimes come to me, confused and frustrated: "Why aren't some patients happy to hear that they are fine? If someone is afraid that they have cancer, and their doctor tells them that the tests show they don't, they will jump up and dance out of the office. But some of my patients look almost disappointed that their tests are normal. They ask, 'But then what *is* wrong with me?' Nothing! Nothing is wrong with them. It's good news—they're fine! But they are still worried, and I have nothing to offer them, because they don't have a problem."

That is where we disagree. These patients are *not* fine, and they know it. They *do* have a problem. There *is*, indeed, something wrong.

"I Can't Think"

Cognitive complaints—the subjective experience of having problems with concentration, learning, memory, word-finding, orientation, problem-solving, organization, or other aspects of thinking—are common throughout the lifespan, although estimates vary widely. In studies of neurologically healthy adults spanning young adulthood to older age, anywhere from 10 to 75 percent endorse cognitive complaints.[1], [2], [3] Overall, it is estimated that about a third of the general population reports cognitive difficulties.[4] And these numbers might be increasing, due in part to successful public education and increased awareness of brain health and conditions like dementia.

Cognitive complaints can be a cause for concern, but they not always are. In some cases, a person's subjective report of cognitive decline, most commonly worsening memory, is the first sign of a condition affecting brain functioning, so cognitive concerns should be taken seriously.[5],[6] This is particularly the case for older individuals, who are at higher risk for certain neurological diseases affecting cognition. However, even among people who are referred to specialty memory clinics for evaluation, in as many as 50 percent of cases no neurological explanation is found for their cognitive lapses.[7] This is consistent with my experience: About one-third of the patients I see in my clinic show no evidence of cognitive deficits on comprehensive neuropsychological testing, despite their reports of significant difficulties in daily life.

Why do these patients feel they "can't remember," "can't think," and "can't function," if their brains are healthy? One answer I encounter repeatedly, and the one I focus on in this book, is that their difficulties are due to cognitive overload created by their unmanageable, unsustainable lives.

This Book

The premise of this book is that many of us are living brain-unfriendly (if not outright brain-hostile) lives. Many of us are going through life chronically stressed, sleep deprived, and either overmedicated or with untreated physical or emotional problems. Psychological symptoms like depression and anxiety, medical conditions like heart disease and chronic pain, the side effects of prescription medications, sleep disorders like insomnia and sleep apnea, the over-use of substances like alcohol and cannabis, the normal changes of aging, and stressful life circumstances related to work, finances, or caregiving can all drain our cognitive resources, effectively reducing the capacity of our otherwise healthy brains.

At the same time, our brains did not evolve to function in a society like ours. We are flooded with endless news to read, notifications to click on, posts to "like," activities to sign the kids up for, medical conditions to "ask your doctor about," and goods to buy. Technology has brought the entire world to the palms of our hands, while our support systems are spread out farther and wider than ever. The volume and pace of information to be processed can overwhelm our cognitive resources.

This perfect storm of brain-unfriendly circumstances—depleted internal resources and cognitive overload from excessive external demands—is the reason many of us find ourselves constantly "glitching": We find musty clothes in the dryer, the overdue work report in our "Drafts" folder, and our car keys in the fridge.

This book presents strategies to help us function better in such a brain-unfriendly world. Part I reviews how our brains process information and the ways in which our modern lives can be a bad fit for it, exhausting even healthy brains and causing them to glitch. Part II focuses on the first component of a brain-friendly life: tending to our depleted brains by addressing conditions that can cause cognitive problems even in the absence of brain pathology by draining our brain's resources. Part III delves into the second component of a brain-friendly life: making our days more brain-friendly by implementing concrete strategies to decrease cognitive overload.

There are a few things you need to know about this book:

First, this is not a book on how to prevent cognitive decline due to aging or dementia. Many excellent books provide evidence-based advice on how to improve long-term brain health and prevent decline. The advice is, appropriately, centered on physical activity, sleep, nutrition, social connection, and mental stimulation. This is not one of those books. Part II does provide recommendations on ways to unburden our brain and optimize its functioning, but the goal of this book is not to offer a set of habits that, if implemented and sustained, will protect brain health over time. Rather, it is to provide strategies, presented in Part III, that we can try *today* as we navigate a brain-unfriendly world—whether or not we also work out regularly, get eight hours of sleep, eat lots of leafy greens, and do daily Sudoku. This book is not about the future, but about helping us be better *now*.

Second, there are books that promise to teach us how to change how our brains work through "rewiring" or "neurohacking." This is not one of those books, either. The premise of this book is that, for many of us, the source of the problem largely resides in how our lives are structured, and not in our brains. The solution I propose is not to mold our brains to survive in an unhealthy reality, but to re-shape our reality to be friendlier to our brains. In other words, the goal of the strategies I present is not to change our brains but to create the conditions that allow them to function better, especially when depleted.

Finally, there are many other books written by experts that review the neuroscientific literature and explain what it reveals about our brain's functioning and the factors that hinder or enhance it. I have a Ph.D. in clinical neuropsychology, and I could have certainly written such a book, but that is not quite the book you are holding. I have worked primarily as a clinician, meaning that I regularly translate the findings from highly controlled research into concrete applications that real patients can implement in their messy worlds. Yes, the advice presented in this book is based on current findings from applied neuropsychological research, but also on my clinical experience working with a widely diverse group of patients over the last 20+ years, and what I have learned from their life experiences.

More importantly, I am also a middle-aged single mother working two jobs and caring for an aging parent long-distance, and that experience informs my views as much as my scientific and professional training. I have written the book I needed when I was a sleep-deprived mom, overwhelmed by responsibility, and frozen at an intersection because I could not remember which way I had to turn to get to my son's friend's house to pick him up, even though I had been there at least ten times, including that morning to drop him off.

Who This Book Is For

This book is written for anyone experiencing cognitive lapses that are not due to brain disease but to a combination of health-related factors decreasing their cognitive bandwidth and excessive demands on their time, energy, and attention. This includes:

- those struggling to adjust to the cognitive changes associated with normal aging;
- those who are experiencing cognitive changes due to a psychological condition like depression or post-traumatic stress disorder;
- those who are experiencing cognitive changes due to a chronic condition like insomnia or fibromyalgia;
- those experiencing the cognitive side effects of a medical treatment, like chemotherapy;
- those who are healthy but still struggling to cope with the many demands related to their work, parenting, caregiving, and other life roles.

Maybe you are concerned enough that you already sought an evaluation, and you were told there is no neurological explanation for your difficulties.

Maybe you are not concerned enough to seek professional help, but you simply want to function better in daily life. This book is for you.

Those living with neurological conditions—like Parkinson's disease or multiple sclerosis, for example—that are causing *mild* cognitive problems might also benefit from the strategies and recommendations presented here. However, this book is not intended for those who have already been diagnosed with a neurocognitive disorder such as dementia, or who are recovering from a neurological event with serious cognitive impact, like a stroke or traumatic brain injury.

Seeking Help

One last but very important point: This book is not meant to discourage anyone concerned about their cognitive lapses from seeking professional help. As I mentioned before, cognitive complaints should be taken seriously because they can be a symptom of a condition needing medical evaluation and treatment. The Appendix offers guidance on how to seek evaluation for cognitive concerns and information about what to expect through that process.

References

1. Blackburn, D.J., Wakefield, S., Shanks, M.F., Harkness, K., Reuber, M., & Venneri, A. (2014). Memory difficulties are not always a sign of incipient dementia: A review of the possible causes of loss of memory efficiency. *British Medical Bulletin*, 112, 71–81.
2. U.S. Preventive Services Task Force. (2020). Screening for cognitive impairment in older adults. US Preventive Services Task Force recommendation statement. *Journal of the American Medical Association*, 323, 757–763.
3. Wooten, K.G., McGuire, L.C., Olivari, B.S., Jackson, E.M., & Croft, J.B. (2023). Racial and ethnic differences in subjective cognitive decline—United States, 2015–2020. *Centers for Disease Control and Prevention Morbidity and Mortality Weekly Report*, 72, 249–255.
4. McWhirter, L.et al. (2020). Functional cognitive disorders: A systematic review. *Lancet Psychiatry*, 7, 191–207.
5. Jessen, F., Amariglio, R.E., van Boxtel, M., Breteler, M., Ceccaldi, M., Chételat, G., et al. (2014). A conceptual framework for research on subjective cognitive decline in preclinical Alzheimer's disease. *Alzheimer's & Dementia*, 10, 844–852.
6. Chapman, S., Rentería, M.A., Dworkin, J.D., Garriga, S.M., Barker, M.S., Avila-Rieger, J., Gonzalez, C., Joyce, J.L., Vonk, J.M.J., Soto, E., Manly, J.J., Brickman, A. M., Mayeux, R.P., & Cosentino, S.A. (2023). Association of subjective cognitive decline with progression to dementia in a cognitively unimpaired multiracial community sample. *Neurology*, 100, e1020–e1027.
7. Ball, H.A., McWhirter, L., Ballard, C., Bhome, R., Blackburn, D.J., Edwards, M.J., Fleming, S.M., Gox, N.C., Howard, R., Huntley, J., Isaacs, J.D., Larner, A.J., Nicholson, T.R., Pennington, C.M., Poole, N., Price, G., Price, J.P., Reuber, M., Ritchie, C., … Carson, A.J. (2020). Functional cognitive disorder: Dementia's blind spot. *Brain*, 143, 2895–2903.

Part I

Why We Glitch

Those of us feeling like we are constantly cleaning up after our brain's mistakes can develop a tense relationship with the organ upstairs. Why can't it just keep up, do better, and stop glitching? A couples therapist often begins treatment by examining a couple's "origin story" instead of jumping into solutions. Similarly, before diving into specific ways to start building a brain-friendly life, we will start by becoming reacquainted with our brain and its workings in Chapter 1. Next, we will review *neurocognitive disorders*— conditions in which cognitive symptoms are caused by diseases affecting the functioning of the brain's cognitive systems. Many of us experiencing lapses worry that they might be signs of such a disease: Have we had a minor stroke? Is this because of that head injury we had years ago in a car accident? Is this early Alzheimer's disease? Chapter 2 will review what the symptoms of such conditions actually look like. Finally, Chapter 3 will elaborate further on my view of the problem presented in the Introduction— that many of us glitch because of the combined effects of living with conditions that deplete our brain's resources and living lives that are in many ways incompatible with how our brains naturally process information. The rest of the book will then focus on the two-pronged answer: how to manage the conditions that drain our cognitive resources and how to make our days a better fit for our brains.

DOI: 10.4324/9781003409311-2

1   How Our Brain Works

Think about everything you might do on a regular weekday morning before you even leave the house. Perhaps it looks a little like this:

> You wake up, likely to an alarm, and wobble to the kitchen to make coffee. You sip it while reading the news on your phone, and as you become more alert, you start thinking about the day ahead: You have to give a committee update at the staff meeting first thing in the morning, and you have to pick up a prescription after work. After showering and dressing, you wake up the kids, feed the dog, and start making breakfast and packing lunches. You remember that your sixth-grader is staying after school for the musical rehearsal, so you throw a couple of extra snacks in their lunchbox. While the kids eat breakfast and argue about who cheated on videogames last night, you go over your presentation in your head. In the middle of all this, you get an automated text reminder with check-in information for your upcoming blood test appointment. You had completely forgotten about it, but you quickly figure out what to do: You will call your coworker to ask him to get the staff meeting started and discuss other agenda items first, since you will now get there late; you will drop your sixth-grader off first, because the lab is closer to your third-grader's school; you will get the test done and then head to work. While you talk to your coworker on the phone, you email him your presentation slides so he can have them ready for you, you pack your workbag, and you signal to your child—twice—that the dog is licking his plate while he argues with his sister. You hang up, tell the kids you're leaving in ten minutes, load the dishwasher, and start looking for your keys.

Even the most typical moments in our lives involve a vast number of mental tasks, from the relatively simple to the incredibly complex. Before talking about what helps our brains function better, it is important to take a look at the brain itself, and how it accomplishes everything it does.

DOI: 10.4324/9781003409311-3

What Is a Brain, Anyway?

The energy costs of the human brain are disproportionally high relative to other organs and to that of other species.[1] Why have such a large, metabolically expensive organ? As human bodies evolved, a larger and faster brain was required to administer the biological resources of such a complex organism, to keep it safe from external threats and meet its internal needs in the most efficient way.[2] Of the many things our brains do to keep us alive and well, we are particularly concerned with cognitive functions organized and executed by the two cerebral hemispheres. Figure 1.1 shows their gross anatomy and main divisions.

It is important to highlight a few features of the brain's design and functioning. First, our brain does not simply *respond* to threats in the environment or to our body's needs. Our brains actually *anticipate* potential threats and needs. Humans with purely reactive brains, who wandered around without concern for predators or awareness that they would eventually get hungry, thirsty, and tired, were unlikely to survive. Our brains are constantly perceiving and processing information from the environment around us and from within our own bodies, learning from experience what helps and what

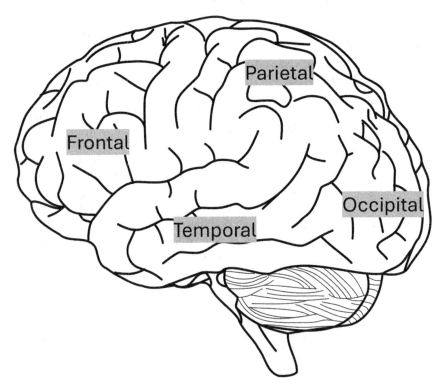

Figure 1.1 The four lobes of the cerebral hemispheres
Source: Unknown, adapted from original.[3]

hurts, and developing flexible behavioral strategies to try to ensure we are safe and our needs are met.

Another feature of the efficiency-focused design of our brains is that it is organized into *networks*. There is no such thing as a single brain region for attention, another for language, another for memory, etc. Instead, we have

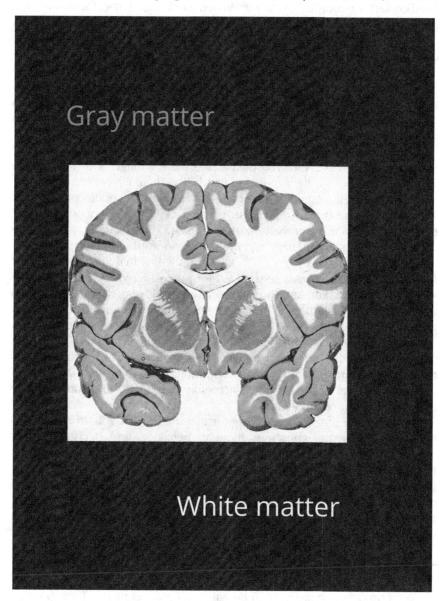

Figure 1.2 Gray and white matter in the cerebral hemispheres
Source: Image by JonesChristiana.[4]

attention networks, language networks, memory networks, emotional regulation networks, and so on, and these networks overlap, allowing them to share brain "real estate" and resources.

To simplify, we can think of these networks as consisting of centers specialized to process certain kinds of information, connected to other specialized centers for coordinated action. In general, these centers for information processing are located in what we call *gray matter*, the areas of the brain that contain the bodies of neurons. Gray matter is found in the *cortex* (the outermost layer of the brain) and in *subcortical* areas deeper in the brain. Communication between centers happens through what we call *white matter*, bundles of *axons*, the fibers that leave the bodies of the neurons and travel short and long distances across the brain to transmit information between different areas. Figure 1.2 shows how gray and white matter are organized in the two cerebral hemispheres.

Think about this simple mental task: You are eating breakfast with your partner and they ask, "Can you please pass me my mug?" It might feel like your hand reaches for their mug and hands it to them almost automatically, without much deliberate thought or effort. Yet something quite complex had to happen in your brain's networks: Areas specialized for processing sound perceived the sounds from your partner's voice. Those sounds were sent to areas specialized for language comprehension for further processing, so you understood "you," "pass," "my," "mug," and the fact that this was a request requiring a response. Areas specialized for processing visual information in space coordinated with areas specialized for eye movements to quickly scan the table for the right object, and areas specialized for object identification recognized the mug. Those visual and spatial areas communicated with motor areas to guide your hand to the mug, position your hand so you could grab it, and guide your hand to where your partner was. Such a simple action engaged multiple networks. Many centers had to process information and communicate with each other incredibly fast for you to complete a task that might feel quite automatic.

A third feature of the brain's design is that the brain is able to *automatize* behavior. Deliberate, conscious thought, behavior, and decision-making require more cognitive resources, more extensive brain activation, and thus more energy. Any activity that can be automatized, performed with little effortful control, saves resources. The classic example is realizing that you drove to a familiar location "without thinking," despite the incredible complexity of the task—operating the vehicle, following traffic signals, navigating the route, etc.

What Brains Do for Us Every Day

For our purposes, we are concerned with the complex mental functions that allow us to flexibly adapt our behavior to our dynamic world so we can achieve our goals, whether our goal is to take a shower or write code for new software. *Cognition*, or *cognitive functions*, refers to this group of complex, higher-order mental processes that allow us to organize our behavior according to situational demands, the environment, and our goals and priorities.[5],[6]

Attention

In its most basic form, attention refers to our brain's ability to focus or to select, out of an environment full of stimuli, what to process (in this book, I will use the term *stimulus* to refer specifically to anything we can perceive through our senses, from the music playing in the background of a department store to a car suddenly cutting us off in traffic or a whiff of our grandmother's perfume). It is the ability to direct our mental spotlight to a particular item or task out of many, so we can engage with it and dedicate cognitive resources to it.

But attention actually involves much more than that. *Sustained attention* refers to the ability to maintain that focus—for example, during a repetitive or ongoing task, like keeping your attention on the road during a long, boring drive. *Selective attention* refers to the ability to focus on certain stimuli while ignoring distracting, competing stimuli—like focusing on the abrupt movements of two cars driving aggressively in front of you, while ignoring your children's bickering in the back seat. *Divided attention* refers to the ability to devote cognitive resources to multiple tasks at the same time—for example, talking on the phone on your car's hands-free device while you try to find an address. True divided attention is rare and difficult. It usually involves two tasks that tap into different systems in the brain, and it works best when at least one of the tasks is relatively automatic. For example, you might be able to carry a conversation while crocheting or playing golf, but even in those cases you might stop talking for a few seconds while you focus on a change in stitch pattern or as you prepare your shot.

Attention relies on broad networks involving parietal and frontal areas, among others. There are other, more complex aspects of attention that we will review below under *Executive Functions*.

Memory

Memory refers to a group of mental processes that allow us to learn, retain, and recall information, or recognize it as familiar. There are many types of memory. *Episodic memory* refers to memory for personal events, like the details you recall from your favorite summer in camp, your first day at your job, or your medical appointment from last week. *Semantic memory* refers to general knowledge of facts and concepts—for example, knowing that World War I started after the assassination of Archduke Franz Ferdinand, that budgies are Australian parakeets that weigh about 30 grams, and what *love*, *all*, and *deuce* mean in tennis. *Prospective memory* refers to the ability to remember to do something in the future, either at a specific time (e.g., take a pill at 11 a.m.) or in response to a cue or event (e.g., ask your physician for a prescription refill next time you see them). More than half of memory lapses that people report in daily life involve prospective memory failures, meaning forgetting to do things they intended to do.[7]

Unlike those kinds of so-called "explicit" types of memory, for information that we can consciously access and verbalize, there are "implicit" types of memory, for information that can be difficult to verbalize and that can even be unconscious. For example, *procedural memory* refers to our memory for complex learned activities, like how to play the guitar, drive a stick shift, or suture a wound. Through *associative learning*, we can learn—sometimes inaccurately and unconsciously—that two stimuli are related. This is why a person with post-traumatic stress disorder (PTSD) caused by a life-threatening car accident might find themselves having a panic attack during a lunch date, unaware that the song playing at the restaurant is the one that was playing on the radio at the time of the crash.

Many structures critical for memory functioning, including the *hippocampi*, are in the temporal lobes, but memory networks extend to include frontal and subcortical areas.

Executive Functions

This term refers to a very broad and varied group of mental processes that help us manage and organize other cognitive resources to make complex goal-directed behavior possible. Executive functions are at the core of what allows us to adapt: Rather than just reacting to what is happening in the environment—sometimes referred to as *bottom-up processing*—executive systems implement *top-down control* of our cognitive resources based on our goals and the situation. Like an orchestra conductor setting the pace and volume of different sections, executive systems can enhance one cognitive process and inhibit another, depending on what needs to be prioritized at any given time to achieve our goals in that particular situation.

Within the umbrella of executive functions we include complex aspects of attention and attentional control. *Cognitive inhibition*, for example, refers to the ability to suppress distractions or irrelevant responses in order to focus on the task at hand. If you are trying to get work done in a busy coffee house, your attention might be pulled by the many distractions in the environment—baristas talking, a cute dog sleeping under a table, cars honking on the street. These are instances of bottom-up processing, when what you focus your attention on is driven by the stimuli themselves—the voices, the dog, the honks. In order to focus on your work, you have to exert top-down control. Based on your goal of getting your work done, you inhibit the processing of those stimuli to instead focus your attention on your paperwork and your laptop screen.

Attentional set-shifting refers to our ability to flexibly shift our attention and cognitive resources between tasks. This is the cognitive ability behind what we call multitasking. We think of multitasking as doing more than one thing at the same time, but in fact what we do is quickly shift our attention back and forth and back and forth between tasks. Remember I said above that true divided attention is rare—most often what we are doing is shifting our attention over and over, extremely quickly, between two things: If you are carefully writing an email and your phone rings, you have to disengage your attention from the email, focus on the phone conversation, then hang up and bring your attention back to where you were on the email. As we will see, this shifting is costly—it takes time and cognitive energy to return to a mental task we interrupted. The cost of shifting will vary depending on how complex or effortful the tasks are. It is easier to handle repeated interruptions, for example, while whipping up a simple dinner than when we are flipping through financial documents to fill our tax forms.

Executive functions are closely related to some aspects of memory. *Working memory*, sometimes considered a type of memory and sometimes an executive function, refers to the ability to briefly keep information in mind and mentally manipulate it while we focus on a task—for example, doing mental math. We rely on working memory whenever we do not write something down, like the four items we need to get at the grocery store, or the three questions we want to ask our child's teacher at a parent–teacher conference. Executive systems also provide critical strategic support for normal memory functioning by helping organize the information we have to learn. To use a metaphor, when executive functions are strong, our memory "archives" are organized, labeled, and color-coded, so it is easy to find, or recall, information. When executive functions are weak, our memory "archives" are like the big kitchen drawer where we throw random things, making it difficult to recall anything at a later time. Executive functions are also critical for prospective memory, so those very common prospective memory lapses we described above are in fact due to executive glitches, not a memory problem per se.

Planning is another executive function. From multiple errands on your way home to a weeks-long trip, planning involves a series of calculations (e. g., about how long each errand will take) and devising logical steps to achieve a certain goal. *Self-monitoring* allows us to check how we are doing in achieving a goal: Whether the goal is to run a half-marathon or getting everyone out of the house on time in the morning, self-monitoring allows us to evaluate whether we are still on track to achieve the goal, or whether we have deviated and need to make a correction to the plan. Planning and self-monitoring allow us to delay gratification in the service of a valued but distant goal—for example, get out of bed and exercise when we would rather lounge, or save money instead of spending it so we can buy plane tickets to go home for the holidays.

While this kind of executive control allows us to follow a plan to get things done, another executive function, *cognitive flexibility*, refers to the ability to adapt to changes in situational demands. Think about having to very quickly come up with a plan when your car stalls on the way to the airport, or when the movers do not show up on the day you have to vacate your apartment. Going through life successfully requires both the ability to plan and stick to a plan *and* the ability to flexibly adapt and problem-solve when circumstances change.

Reasoning and judgment are also executive functions, referring to the ability to reach rational conclusions and make sensible decisions to achieve our goals given the circumstances. They refer to the ability, for example, to decide between having surgery or not based on information about the rates of success, the likely outcomes, and the likelihood of different complications. They are the reason we do not respond to an email telling us we won a large sum of money that will be deposited into our bank account as soon as we reply with our account information.

A final group of executive functions is related to behavioral and *emotional regulation*. As highly social animals that largely depend on others for survival, our ability to develop and maintain close social relationships is crucial. Our brains automatically process social information, like other people's facial expressions and body posture. Executive systems allow us to interpret others' emotions and behaviors, like picking up on disingenuous expressions of interest, knowing when someone wants to be left alone, or realizing our joke caused offense instead of amusement. They also allow us to express our anger, frustration, and disappointment in socially appropriate manners—for example, to say "Thank you for the feedback" when our boss unfairly and harshly criticizes us, and to speak calmly and lovingly when our toddler has their fourth meltdown of the day.

The frontal lobes and prefrontal areas in particular (the areas at the very front of the brain) contain structures critical for executive functions. These executive networks have dense connections with subcortical and parietal areas. You might have heard about the *limbic system*, a group of structures, including the *amygdala*, that are important for the processing of emotional and other salient stimuli, which can happen without our control and even outside of our awareness—think about tearing up unexpectedly when you hear your late mother's favorite song, or disliking someone "for no reason" because you are unaware that they look like the babysitter that used to scare you with horror stories. Not surprisingly, some structures that are part of memory systems are also part of the limbic system, revealing a close connection between our memory and our emotions. Some executive networks have close connections with limbic structures and are responsible for regulating (not always successfully) how we process and respond to emotional information.

Language

In broad terms, there are *expressive* language functions, involving our ability to convey our thoughts by producing the right words linked together into coherent sentences, and *receptive* language functions, our ability to under-stand what we hear and read. During verbal interactions, we rely on multi-ple language functions, including automatic verbal production ("I know, right?" "Not at all," "Bless you!"), more deliberate articulation of thoughts, comprehension of what others are saying, and processing of nonverbal aspects of communication like the tone of the other person's voice (was that a sincere expression of concern or a sarcastic dig?). Language functions rely on complex networks involving temporal areas specialized in language comprehension, and frontal areas specialized for language production.

Visuoperceptual and Visuospatial Functions

There are several perceptual systems in the brain dedicated to the proces-sing of incoming information through every sensory channel, as well as internal information about the position and movement of our own bodies, referred to as *proprioception*. For our purposes, it is important to highlight that we have visual perceptual systems devoted to identifying objects and faces, like when you pick the right tool out of a packed toolbox or find your child's face in the chaos of the school holiday play. We also have *visuos-patial* systems, which allow us to perceive the location of objects and navi-gate space—for example, when we quickly visualize a new route to get to work on time after encountering a street closure.

The occipital lobes process basic visual information. Object and face recognition relies on networks involving occipital and temporal areas, and visuospatial processing relies on networks involving occipital and parietal areas.

Motor Functions

There are many motor systems that carry out automatic and intentional, simple and complex movements, including the ability to get dressed, use a screwdriver, wrap a burrito, and use home appliances. Motor systems rely on networks involving subcortical structures and areas in the frontal lobes, in close connection with parietal visuospatial systems that guide our move-ments in space. Some more automatic motor sequences, like filling our car with gas or throwing a baseball back and forth, can be handled without much effortful control and without interfering with other activities, like car-rying a conversation. Other motor tasks, like putting together a new piece of furniture, require more attentional and executive resources, and are more difficult to execute while engaged in other tasks.

Cognitive Processing Speed

Finally, in order for us to be able to make it through our days successfully, our brain has to be able to do all of this at a certain speed. Cognitive processing speed refers to how fast we are able to process internal and external information—for example, whether we can keep up with our young server as he peevishly recites tonight's specials, how quickly we can do mental math to figure out if we got the correct amount of change at a very busy food stand, and how quickly we can come up with an excuse for why we can't stay late when our boss puts us on the spot in the middle of a meeting. Cognitive processing speed relies on different information-processing centers being able to communicate quickly and efficiently. Because of this, the integrity of white matter tracts is crucial for adequate cognitive speed.

* * *

Why would these awe-inspiring devices between our ears malfunction? There are many avenues to cognitive dysfunction. Before we delve into how cognitive lapses can occur in the absence of brain disease, due to conditions that deplete the cognitive resources of a healthy brain or due to excessive life demands, it is important to review the cognitive problems that occur due to neurological diseases that directly affect brain structure and functioning. We will talk about such neurocognitive disorders next.

Resources

- www.BrainFacts.org is an initiative of the Society for Neuroscience, the world's largest organization of brain scientists. This site presents information about the brain for the general public, including information about brain functioning, neurological conditions, and the application of neuroscience to social issues like law, art, and technology.
- The National Academy of Neuropsychology has a website, www.BrainWiseMedia.com, dedicated to public education about brain health and disease, and a podcast, BrainBeat, featuring interviews with experts.
- If you are interested in the fascinating topic of the evolution of the human brain, an excellent source is the book *What Is Health? Allostasis and the Evolution of Human Design* by Peter Sterling, a concise but in-depth account of the biological bases of human neural design.
- For a comprehensive and engaging account of memory functions and their fallibility, the book *The Seven Sins of Memory, Updated Edition: How the Mind Forgets and Remembers* by Dan Schacter reviews decades of research on human memory systems from the fields of neuroscience and neuropsychology. You can also listen to his interview in the American Psychological Association's podcast *Speaking of Psychology*, episode 158 (www.apa.org/news/podcasts/speaking-of-psychology).

References

1. Kuzawa, C.W., Chugani, H.T., Grossman, L.I., Lipovich, L., Muzik, O., Hof, P.R., Wildman, D.E., Sherwood, C.C., Leonard, W.R., & Lange, L. (2014). Metabolic costs and evolutionary implications of human brain development. *PNAS*, 111(36), Article 13011.
2. Sterling, P. (2020). *What is health? Allostasis and the evolution of human design.* MIT Press.
3. Unknown, edited by Amousey (2020). Brain Outline Lateral [Image]. Wikimedia Commons. https://commons.wikimedia.org/wiki/File:Brain-outline-lateral.svg.
4. JonesChristiana (2022, June 13). Gray and white matter of the cerebrum [Image]. *Wikimedia Commons*. https://commons.wikimedia.org/wiki/File:Gray_and_White_matter_of_the_cerebrum.png. License information: CC BY-SA 4.0. https://creativecommons.org/licenses/by-sa/4.0. (Original file.)
5. Loring, D.W. (Ed.). (2015). *INS dictionary of neuropsychology and clinical neurosciences* (2nd ed.). Oxford University Press.
6. McFarland, D.J. (2017). How neuroscience can inform the study of individual differences in cognitive abilities. *Reviews in the Neurosciences*, 28(4), 343–362.
7. Leong, R.L.F., Cheng, G.H.-L., Chee, M.W.K., & Lo, J.C. (2019). The effects of sleep on prospective memory: A systematic review and meta-analysis. *Sleep Medicine Reviews*, 47, 18–27.

2   Neurocognitive Disorders

I am writing this chapter on a sunny winter morning in Tucson, Arizona, where on January 8 of 2011 a gunman opened fire at a "Congress on Your Corner" event, killing six people and injuring 13 others, including U.S. Representative Gabrielle Giffords. In the moments after being shot in the head at point-blank range, Gabby was able to understand and follow commands—for example, squeezing the fingers of a paramedic with her left hand. However, she was unable to move the right side of her body and she was unable to speak. During her acute recovery period, she was only able to say a few words, and she at times perseverated on words unrelated to what was being talked about. In their memoir, her husband, former astronaut and now U.S. Senator Mark Kelly, describes a visit former President George H. W. Bush paid Gabby at the Houston hospital where she was undergoing rehabilitation, about two months after her injury:[1]

> "You've been so strong," he told her. "I'm really proud of you. And I'm praying for your recovery."
> "Chicken," Gabby replied. "Chicken."

* * *

A patient recovering from a large brain hemorrhage completes his neuropsychological evaluation on a Monday, during which he denies any cognitive problems, then shows up again on Tuesday and again the following week, insisting that he has an appointment to be evaluated. A retired English teacher recovering from an ischemic stroke cannot select from a display the picture that shows a "carrot" or a "cookie," and cannot name any animals other than "dog." A young patient who stopped breathing for several minutes due to a drug overdose leaves the office for a lunch break, eats her lunch at the café downstairs, then goes home, forgetting she was supposed to come back upstairs to finish her testing. When I show a patient with moderate Alzheimer's dementia a comb, he tells me he does not know what that is and that he has never seen such an object before, but later absent-mindedly combs his hair with it. An older patient with frontotemporal

DOI: 10.4324/9781003409311-4

dementia tells a moving tale of being a nurse during World War II, but her daughter, sitting behind her, shakes her head and mouths that none of it is true. A middle-aged business manager recovering from a serious brain infection meticulously replies to every spam email and gives out her social security number and mother's maiden name, resulting in thousands of dollars of fraudulent charges to her credit card. A young man survives a serious head injury from a car accident with no obvious cognitive impairments, but displays such severe impulsivity and poor decision-making that he is unable to keep a job.

All of these individuals were living with profound cognitive impairments caused by conditions that damaged structures in their brains or otherwise disrupted the functioning of brain networks critical for cognition. In this chapter, we will review the nature of cognitive symptoms seen in some common neurological disorders.

Cognitive Symptoms and Neurocognitive Disorders

A cognitive *symptom* refers to a reported difficulty with some aspect of higher-order mental processing. The problem represents a change—a decline—from the person's baseline level of functioning, and it is noticeable to the individual experiencing it or to a knowledgeable observer, like a close family member or an attentive healthcare provider.

Neurocognitive disorders refer to disorders characterized primarily by the presence of cognitive symptoms that affect how the person functions in daily life.[2] The cognitive symptoms are caused by a known or suspected neuropathology or medical condition affecting the brain's normal functioning: Alzheimer's disease, stroke (also referred to as a cerebrovascular accident, and which can be either *ischemic*, when it is caused by blockage in a blood vessel, or *hemorrhagic*, when it is caused by bleeding from a ruptured blood vessel), traumatic brain injury (TBI), Lewy bodies disease, epilepsy, Parkinson's disease, chronic alcoholism, multiple sclerosis, brain infections, lack of oxygen to the brain (e.g., from cardiac arrest), and many other conditions that affect the structural integrity or functioning of brain networks can cause a neurocognitive disorder.

A few things are important to note. First, we mentioned in Chapter 1 that the brain is organized in networks, with processing centers and connections between these centers. By observing cognitive symptoms, we can infer which brain network is affected. For example, Gabby Giffords being able to understand language but not speak indicated that the trajectory of the bullet had affected the frontal networks for language production, but had spared the temporal networks for language comprehension.

Second, the symptoms do not necessarily tell us what specific disease or neuropathology is causing them. In Gabby's case, the reason for the injury was obvious. But if an older adult presents to the emergency department unable to speak, we cannot know from the symptom alone whether the cause is a stroke, seizure activity, or a type of primary progressive aphasia, a dementia affecting expressive language. Different diseases can cause the same symptom by affecting the same networks through different pathologies and mechanisms.

Third, some of the diseases that cause neurocognitive disorders are *neurodegenerative*, meaning they have a progressively worsening course, while other conditions can remain stable or even improve to some degree. For example, Alzheimer's disease is a neurodegenerative disease with—as of now—no cure, so the cognitive symptoms gradually worsen through the end of life. In contrast, someone recovering from a TBI or a stroke might experience significant improvement in the months or even years after the event—Gabby Giffords regained some speech and movement—before reaching a plateau and remaining stable in the absence of further neurological insult.

How Are Neurocognitive Disorders Diagnosed?

The diagnosis of a neurocognitive disorder is based on (a) how severe the cognitive impairments are, as documented on formal neuropsychological testing, and (b) to what extent the cognitive impairments affect the person's ability to perform *activities of daily living*, which include complex activities such as managing finances and medications, and more basic activities like preparing meals and self-care.

The language used in the diagnosis of neurocognitive disorders can be confusing for patients, because different healthcare professionals use different terms or mean different things by the same terms, depending on their background and the specific diagnostic system they use in their practice. The Appendix provides more detailed information on the diagnostic process and distinctions. Broadly, the terms *mild cognitive impairment* (often abbreviated "MCI") and *mild neurocognitive disorder* refer to cognitive symptoms that, while noticeable, are relatively modest and do not interfere with the person's ability to perform activities of daily life independently, although it might be more difficult and require more time, effort, and supports. For example, a person with MCI after a TBI might be able to return to their office job with supports including an electronic calendar with automated reminders for deadlines, an early morning "huddle" with their supervisor to plan priorities for the day, and multiple breaks to help them shift from task to task.

The terms *dementia* and *major neurocognitive disorder* refer to cognitive symptoms that are severe enough that the person is unable to function independently, even with supports like reminders and alarms. For example, if you have a loved one with dementia, you might write the day and time of a medical appointment on their calendar and you might also call them that morning to remind them that you are picking them up in about an hour to take them to their appointment, but by the time you get there, they might be surprised to see you and not recall what they were supposed to do or where they are going.

The diagnosis of a neurocognitive disorder also involves determining the underlying neuropathology responsible for the symptoms. For example, one person's dementia could be caused by Alzheimer's disease, another person's dementia might be caused by multiple strokes, and another person's dementia might be caused by a combination of both pathologies. A person's MCI can be caused by a TBI, while another's might be caused by a mild stroke. The Appendix will provide more details on how these determinations are made.

What Symptoms of Neurocognitive Disorders Look Like

Neurocognitive disorders can affect any of the cognitive functions reviewed in Chapter 1.

Attention

Impairment in basic attention, meaning the ability to focus and maintain attention, is primarily seen in conditions that cause broad brain dysfunction—for example, in individuals who are in a state of delirium due to a serious infection, experiencing intoxication or withdrawal from a psychoactive substance, or in an acute psychiatric episode.[3] In most neurocognitive disorders—and even in a condition like attention-deficit/hyperactivity disorder (ADHD)—impairments are more commonly seen in more complex aspects of attention, described below under *Executive Functions*.[4]

Memory

Many neurocognitive disorders cause some degree of memory impairment, either by disrupting memory networks themselves or by disrupting executive functions, which in turn affects memory functioning. A classic example of *amnesia*, or primary memory impairment resulting in inability to learn, retain, and/or recall information, is seen in patients with Alzheimer's disease. Alzheimer's pathology typically affects memory networks early on, and as a result patients display profound impairment in episodic memory, with inability to learn and retain new information.[5] Because of this, they are often repetitive: You might tell them in the morning that you are meeting your neighbors, Bob and Sheila, for lunch, and a few minutes later, they might ask you what you have planned for the day. The next day, or the next week, they might not recall going out for lunch. Because this is a true, primary amnestic disorder, prompts and cues do not help. It is not helpful to prod a patient with Alzheimer's ("Don't you remember? Remember we were with Bob and his wife? Remember her name? It starts with Sh...? Remember what you ordered? You had your favorite..."). The person is not having difficulty retrieving the information—the information is truly forgotten, lost, so there is nothing to retrieve, and prompts and cues often only increase frustration. Initially, remote memories tend to be preserved, but as the disease progresses, the person not only has difficulty acquiring new information, like the name of their new grandchild, but they forget older memories, like the kind of work they used to do, their wedding day, and their semester abroad in college.

Impairment in semantic memory, or the loss of knowledge, is more rare. Alzheimer's patients experience loss of semantic knowledge later in the disease, and in a condition called *semantic dementia*, the most prominent early symptom is the loss of such knowledge.[6] A pastry chef with extensive knowledge of exotic fruits who develops semantic dementia might be stumped when a recipe calls for a kiwi, unable to think of what that is. As the disease progresses, they might only be able to identify even common fruits, like an apple, as "some kind of fruit." Eventually, they will be unable to say what a "fruit" is, give an example of a fruit, or recognize any fruits, because the whole concept of *fruit* has been lost.

Executive Functions

Executive dysfunction is one of the most common symptoms in neurological conditions affecting brain functioning. Individuals who survive moderate to severe TBIs, for example, can experience chronic impairments to a variable degree on any aspects of executive functioning.[7] Deficits in *attentional control* can cause them to jump from one task to the next, leaving them all unfinished. Deficits in *cognitive inhibition* can result in distractibility, because they are unable to stop themselves from responding to whatever is happening around them instead of staying focused on the task at hand. Impairment in *attentional shifting* can result in patients *perseverating* on a topic or behavior. For example, a patient who has not driven since their TBI who is told they could eventually take a test to see if they can start driving again might perseverate heavily on this, making multiple calls and appointments with their doctor to talk about getting their license back, despite being told repeatedly they are not ready. Impaired *reasoning* and judgment can result in unwise decisions (like sending large amounts of money to a stranger they met online) and vulnerability to exploitation. Impaired *regulation* of emotions and behaviors translates into impulsivity, explosive anger, and socially inappropriate behaviors, which can be disabling even in the absence of other impairments.

Many patients with neurological disorders experience memory problems secondary to executive dysfunction. As we discussed in Chapter 1, executive systems help us organize information so it can be more efficiently learned and eventually recalled. Patients with mild cognitive problems due to Parkinson's disease, for example, often experience memory difficulties due to executive dysfunction.[8] Because their problem is not with memory systems themselves but with "executive" or strategic aspects of memory, these patients often benefit from memory aids and cues. Unlike a patient with Alzheimer's disease, information has been learned and retained, but they are having difficulty retrieving it from their disorganized memory "archives," so prompts can help.

You might remember that executive systems are also responsible for self-awareness and monitoring. Because of this, neurological diseases affecting executive networks often cause *anosognosia*, a lack of awareness of the deficits. Patients often display significant cognitive impairments but insist that they are fine and complain that their loved ones are making them undergo the evaluation.[9]

Language

Depending on what specific component of the language network is affected, impairment in language functions—or *aphasia*—can take the form of problems with comprehension (receptive aphasia) or producing language (expressive aphasia). Because of the anatomy and distribution of blood vessels in the brain, language networks are often affected in the most common types of stroke, and thus aphasia is a common sequela of stroke.[10]

Because people with receptive aphasia are unable to understand language, they cannot understand what others say to them, and their own speech can be nonsensical. For example, a patient with profound receptive aphasia, asked what they are doing after the appointment, might very fluently and clearly answer something like "I'm going with the road over there all the time and going with the people for them over there. We're walking with the people all the time, but we'll send all the time help road very soon, for them, we hope, for them."

People with expressive aphasia, on the other hand, are able to understand what others say, but have difficulty expressing themselves through language because of inability to produce words and sentences. A patient with profound expressive aphasia, asked what they are doing after the appointment, might respond, "I'm...food. She...uh...uh...wife, and, uh...my...me...food for me. Uh...uh... The...chil...children...and, uh...four—five. Five, uh...children, and uh...music, and uh...uh, uh...laugh, and, uh...talk."

People with milder language impairments might produce largely normal conversational speech but with word-finding difficulties ("Then we pulled up to the...the...you know, the corner...where the roads meet...the...the intersection!"). They might make errors called *paraphasias*, meaning they produce an incorrect but usually phonologically or semantically related word. For example, a patient wanting to say, "We took the boat out on the lake," might say, "We took the sail out on the lake" or "We took the poat out on the lake." Others display *anomia*, the inability to name known objects shown to them. For example, when shown a picture of a piano, they might say, "Oh, for music, you play it," while moving their fingers as if playing the piano, indicating preserved recognition. If you ask, "Is it a piano?" they immediately say, "Yes! A piano." They have not forgotten what a piano is, they just cannot retrieve the word; it is not a memory problem but a language problem. (In contrast, if you show the pastry chef with semantic dementia an apple, he will not only be unable to name it, but he will say he does not know what that is. If you ask, "Is it an apple?" he might shrug and say, "No idea." His failure to produce the correct name is not simply due to a language problem interfering with word retrieval; in his case, the conceptual knowledge of what an "apple" is has been lost.)

Visuoperceptual and Visuospatial Functions

Disruption to components of the visual perceptual system can result in *agnosia*, or the inability to recognize known objects, like my patient who did not know what he was looking at when I showed him a comb.[11] Notice that in agnosia, the problem is not loss of knowledge or difficulty coming up with the word; the problem is recognizing the object. My patient could not recognize a comb due to his visual perceptual impairment, but if you ask him what a comb is, he would say, "Something you use to fix your hair." He knew what a comb was, but he could not recognize one when he saw it.

Dysfunction in visuospatial systems can result in a variety of symptoms, including *spatial disorientation*. Getting lost in a familiar environment, like driving home from the neighborhood grocery store, is one common reason patients with Alzheimer's disease are brought in for evaluation by their loved ones, who might until then not have realized the extent of their cognitive decline.[12]

Motor Functions

Disruption to certain motor systems can result in *apraxia*, an impairment of learned, skilled movements.[13] Patients who develop apraxia from strokes affecting areas critical for controlled movements, for example, can have difficulty using appliances like a coffee maker, brushing their teeth, or getting dressed—they might pour water into the coffee pot itself instead of the water tank, attempt to brush their teeth by making random circular motions in front of their mouth without putting toothpaste on the toothbrush, and turn their shirt over and over, unable to figure out where their hand should go through.

Processing Speed

Patients with impaired processing speed experience significant slowness across cognitive processes. They take much longer to focus and shift their attention, learn and recall information, make decisions, problem-solve, and think in general. Slow processing speed is one of the most common early cognitive symptoms in Parkinson's disease, likely due to neurotransmitter abnormalities.[14] Slow processing speed is also common in disorders affecting the integrity of the white matter. For example, a traumatic brain injury can cause shearing of the axons that compose the white matter,[15] vascular disease can damage the blood vessels that supply blood to the white matter,[16] and in multiple sclerosis, an autoimmune process causes lesions in *myelin*, a protective sheath that covers axons and increases the speed of signal transmission.[17] Slow processing speed is commonly seen in all of these conditions.

Impaired processing speed often causes impairments in other cognitive domains, because when our brain cannot process information fast enough, we cannot fully pay attention to information, and because of that we cannot remember it at a later time. For the person with impaired processing speed, it can feel like the world is moving as fast as the disclaimers at the end of a medication ad on TV. They miss what people say, which can make it look like they forgot what they were told, and they often feel like they just can't think.

* * *

As fascinating as it can be to delve into neurocognitive disorders, the pathologies behind them, and what they reveal about the workings of the brain, this book is concerned with a different issue: Why people experience cognitive symptoms in the absence of neuropathology. We turn to that topic next.

Resources

- The website for the National Institute of Neurological Disorders and Stroke (www.ninds.nih.gov) has extensive information and resources on a broad range of neurological conditions.
- The American Academy of Neurology has a public website, Brain & Life (www.brainandlife.org), with a digital magazine (a free print version subscription is available) and a podcast on topics related to brain health and neurological conditions.
- For information on specific conditions, reliable sources include the websites for the Alzheimer's Association (www.alz.org), the American Stroke Association (www.stroke.org), the Brain Injury Association of America (www.biausa.org), the Association for Frontotemporal Degeneration (www.theaftd.org), the Parkinson's Foundation (www.parkinson. org), the Michael J. Fox Foundation (www.michaeljfox.org), the National Multiple Sclerosis Society (www.nationalmssociety.org), and the Epilepsy Foundation (www.epilepsy.com).

References

1. Giffords, G., & Kelly, M. (2011). *Gabby: A story of courage and hope*. Scribner.
2. American Psychiatric Association. (2022). Neurocognitive disorders. In *Diagnostic and statistical manual of mental disorders* (5th ed., Text Revision). American Psychiatric Association Publishing.

3. Wilson, J.E., Mart, M.F., Cunningham, C., Shehabi, Y., Girard, T.D., MacLullich, A.M.J., Slooter, A.J.C., & Ely, E.W. (2020). Delirium. *Nature Reviews Disease Primers*, 6, Article 90.
4. Willcut, E.G. (2023). Neuropsychology of attention-deficit/hyperactivity disorder. In G.G. Brown, T.Z. King, K.Y. Haaland, & B. Crosson (Eds.). *APA handbook of neuropsychology Vol. 1. Neurobehavioral disorders and conditions: Accepted science and open questions*. American Psychological Association.
5. Weintraub, S., Wicklund, A.H., & Salmon, D.P. (2012). The neuropsychological profile of Alzheimer disease. *Cold Spring Harbor Perspectives in Medicine*, 2, Article a006171.
6. Gorno-Tempini, M.L., Hillis, A.E., Weintraub, S., Kertesz, A., Mendez, M., Cappa, S.F., Ogar, J.M., Rohrer, J.D., Black, S., Boeve, B.F., Manes, F., Dronkers, N.F., Vandenberghe, R., Rascovsky, K., Patterson, K., Miller, B.L., Knopman, D.S., Hodges, J.R., Mesulam, M.M., & Grossman, M. (2011). Classification of primary progressive aphasia and its variants. *Neurology*, 76, 1006–1014.
7. Wood, R.L., & Worthington, A. (2017). Neurobehavioral abnormalities associated with executive dysfunction after traumatic brain injury. *Frontiers in Behavioral Neuroscience*, 11, Article 195.
8. Dirnberger, G., & Jahanshahi, M. (2013). Executive dysfunction in Parkinson's disease: A review. *Journal of Neuropsychology*, 7, 193–224.
9. Sunderaraman, P., & Cosentino, S. (2017). Integrating the constructs of anosognosia and metacognition: A review of recent findings in dementia. *Current Neurology and Neuroscience Reports*, 17, Article 27.
10. Grossman, M., & Irwin, D.J. (2018). Primary progressive aphasia and stroke aphasia. *Continuum Behavioral Neurology & Psychiatry*, 24(3), 745–767.
11. Bauer, R. (2012). Agnosia. In K.M. Heilman & E. Valenstein (Eds.). *Clinical neuropsychology* (5th ed.). Oxford University Press.
12. Silva, A., & Martínez, M.C. (2023). Spatial memory deficits in Alzheimer's disease and their connection to cognitive maps' formation by place cells and grid cells. *Frontiers in Behavioral Neuroscience*, 16, Article 1082158.
13. Heilman, K.M., & Gonzalez Rothi, L.J. (2012). Apraxia. In K.M. Heilman & E. Valenstein (Eds.). *Clinical neuropsychology* (5th ed.). Oxford University Press.
14. Vriend, C., van Balkom, T.D., van Druningen, C., Klein, M., van der Werf, Y.D., Berendse, H.W., & van den Heuvel, O.A. (2020). Processing speed is related to striatal dopamine transporter availability in Parkinson's disease. *Neuroimage: Clinical*, 26, Article 102257.
15. Marquez de la Plata, C.D., Garces, J., Kojori, E.S., Grinnan, J., Krishnan, K., Pidikiti, R., Spence, J., Devous Sr., M.D., Moore, C., McColl, R., Madden, C., & Diaz-Arrastia, R. (2011). Deficits in functional connectivity of hippocampal and frontal lobe circuits after traumatic axonal injury. *Archives of Neurology*, 68(1), 74–84.
16. Cannistraro, R.J., Badi, M., Eidelman, B.H., Dickson, D.W., Middlebrooks, E.H., & Meschia, J.F. (2019). CNS small vessel disease. *Neurology*, 92, 1146–1156.
17. Lie, I.A., Weeda, M.M., Mattiesing, R.M., Mol, M.A.E., Pouwels, P.J.W., Barkhof, F., Torkildsen, Ø., Bø, L., Myhr, K.-M., & Vrenken, H. (2022). Relationship between white matter lesions and gray matter atrophy in multiple sclerosis: A systematic review. *Neurology*, 98(15), Article e1563.

3   A Brain-Unfriendly Life

On January 29, 2024, the Elmo account on X (formerly Twitter) tweeted, "Elmo is just checking in! How is everybody doing?"[1] The tweet quickly went viral, with over 200 million views and over 20,000 responses. While many comments were warm and funny, some of the most "liked" responses included "Suffering," "Every morning, I cannot wait to go back to sleep," "The world is burning around us, Elmo," and "Elmo, we are tired." The overall tone of the thread was such that the official Sesame Street account eventually replied with a link to mental health resources, and even President Biden chimed in, commenting on the importance of seeking help when needed and offering support to others. The episode was widely discussed in articles with headlines like "Elmo wrote a simple tweet that revealed widespread existential dread,"[2] "Elmo's viral tweet sparks an existential crisis among his followers,"[3] and "The internet is trauma-dumping on Elmo."[4]

Commentary on adult life is a common type of viral social media post, often receiving tens, sometimes hundreds of thousands of "likes" and retweets, and often compiled in online articles that describe them as "hilarious" and "relatable." Some examples: "One of my favorite games to play is 'is my headache from dehydration, caffeine withdrawal, lack of proper nutrition, my ponytail, stress, lack of sleep, not wearing my glasses, or brain tumor?'"[5] "[What] is a 'group chat?' I am an adult, I only have two friends and they don't know each other,"[6] and "Sorry can't come out tonight, I gotta sit alone in front of a muted TV opening and closing the same 4 social media apps until 3am."[7]

A popular meme is the phrase "How your email finds me" accompanied by a picture of a person looking depressed, sick, or engulfed in flames, or of a disheveled, rabid, or semi-conscious animal. Content that makes fun of our reliance on substances is ubiquitous and popular, like a widely shared cartoon (available on mugs and T-shirts) of a cup of coffee and a glass of wine on a running track, with the coffee cup passing the baton to the glass of wine under the heading "Literally every day." One tweet that stood out to me was "Oh boy ever spill a little bit of your coffee and realize the thread you are hanging on by is actually quite thin,"[8] partly because of its date: It

DOI: 10.4324/9781003409311-5

was tweeted in December of 2019, just a couple of months before the Covid pandemic hit.

While much of this content is certainly shared for laughs, I have to say that, as someone who talks to people about their lives for a living, that last tweet captures my impression quite accurately: That many of us were in fact "hanging by a thread" and that the sudden and dramatic disruptions to daily life that the pandemic brought about pushed many of us over the limit. Many of us have a feeling that life as we are living it is unsustainable. We do not take care of our bodies, we do not cultivate relationships, we stare at multiple screens for long periods of time, and we routinely rely on substances like caffeine and alcohol to "make it through the day." Some of the responses to Elmo's well-meaning tweet seemed to reflect a craving for connection and a hunger to say out loud that some of us are not okay.

* * *

How many of these statements apply to you?

- I have 2 or more chronic health conditions.
- I live with chronic pain.
- I take 5 or more medications a day.
- I take pain and/or sleep medications daily.
- I usually do not feel rested when I get out of bed in the morning.
- I usually nod off if I sit on the couch in front of the TV in the afternoon.
- I have more than 2 alcoholic drinks a day.
- I do not feel like I can handle or cope with all my personal responsibilities and problems.
- I do not feel in control of important things in my life.
- I feel nervous, stressed, or anxious.
- I feel sad or down.
- I do not feel like doing anything.
- I do not enjoy doing things like I used to.
- I keep thinking about really bad things that have happened.
- I wish I had someone to turn to for support and to talk to about how I really feel.
- I feel I do not have enough time to do everything I need to do.
- I can rarely do a single thing for more than a few minutes without interruption.

The more statements you endorsed, the more likely it is you are living a "brain-unfriendly" life, with unaddressed issues that are draining your cognitive resources and/or cognitive overload from excessive demands tied to your life roles and responsibilities. But even a single one of these, like living with multiple chronic conditions requiring complex medical treatment, can be enough to deplete your brain and cause it to glitch.

If Brains Are So Amazing, Why Are We Struggling?

Let's first point out that mild cognitive lapses are common in healthy people across the lifespan. Between one-fourth and one-third of adults in the general population report poor concentration, forgetfulness, and slow thinking. [9] Even among young, healthy college athletes, over 10 percent report difficulty remembering and about 15 percent difficulty concentrating.[10]

But why would someone without a brain disease experience cognitive symptoms significant enough to interfere with their daily life and raise concern that maybe they are suffering from a neurological condition? What I have seen over and over, in more than 20 years asking patients in-depth questions about their life histories, is that people walk around carrying unimaginable burdens: Stories of heart-wrenching trauma, more life roles and responsibilities than seems humanly possible to take on, profound socioeconomic deprivation and financial stress, and diseases and surgeries resulting in ongoing pain and disability, among others. Some of them are sitting in my office because they have come to the realization that they cannot quite continue living the same way, while others seem unaware of how truly heavy the loads they carry are.

A Depleted Brain

The first reason for our cognitive lapses is a group of common factors that deplete our brain resources. We are more likely to glitch when we are living with unaddressed psychological issues, unmanaged physical issues, and chronically stressful life circumstances. As I mentioned in the Introduction, all of the following can contribute to cognitive problems by draining our mental resources:

- chronic stress related to multiple life roles and responsibilities, including parenting, caregiving, financial responsibilities, and work;
- mental health conditions like depression, anxiety, post-traumatic stress disorder, bipolar disorder, and psychosis;
- medical conditions, including serious diseases like advanced organ failure (e.g., congestive heart failure, end-stage renal or liver disease) and chronic respiratory conditions, but also not fully controlled chronic conditions like diabetes, hypertension, chronic pain, endocrine disorders, and significant nutritional deficiencies;
- prescribed treatments for some medical conditions, including certain pain, sleep, psychiatric, neurological, and chemotherapy medications;
- sleep disorders like insomnia and sleep apnea, but also mildly but chronically insufficient sleep, or sleep of poor quality;
- regular use of substances with both acute and chronic effects, like alcohol and cannabis, even if their use is not otherwise problematic; and

- changes associated with normal, healthy aging. While it is not really accurate to say that aging "depletes" our brains, it is certainly associated with decreased capacity in many cognitive systems.

There are two important points regarding the impact that these conditions can have on your cognitive functioning: One is that while I'm using terms like "medical conditions" and "mental health" or "psychological conditions," the distinction between "physical" and "mental" is an artificial one. It might be easy to think that the neurological diseases mentioned in Chapter 2 obviously cause cognitive symptoms because they damage or disrupt the functioning of the physical neural networks critical for cognition, but that a "mental" experience like depression or stress could not cause "real" cognitive issues. In fact, over the last decades neuroscience has revealed the neurobiological aspects of experiences we used to think of as psychological. We have learned, for example, of neurochemical abnormalities and abnormal brain activity in depression, chronic stress, and sleep deprivation. "Depression" and "stress" are truly neuro-psychological phenomena; they are not "all in your head" in the figurative sense, but they are, indeed, in your head, affecting how your brain functions.

The other important point is that cognitive *change* does not mean cognitive *impairment*.[11] There is a wide range of what is considered normal functioning, while impairment refers to performance that is significantly below that normal range, rendering the person unable to perform certain cognitive tasks. Many individuals with the conditions listed above can experience a change or decline in their cognitive functioning, while remaining within the "normal" range.

Imagine the memory test shown in Figure 3.1. The test has a score range of 0 to 100, and "normal" scores for adults of a certain age range from 25 to 75 points. Let's say a healthy person typically scores a 70 on this test, as shown by the fully charged battery symbol. Now imagine this person develops chronic pain after hip surgery. The pain disturbs their sleep, so they are prescribed pain and sleep medication, which makes them feel sedated. They also experience depression due to the pain and physical limitations. Under those circumstances, in a depleted state, this person might now score a 45 on the same memory test, as shown by the low-battery symbol. A drop—a decline—from a score of 70 to a score of 45 is quite significant, and the person might notice very real differences and lapses in their daily life. Their brain is clearly not functioning optimally. However, their performance remains in the "normal" range, meaning their memory is not impaired.

Similarly, as we review the impact that these conditions have on cognitive functioning, we will often refer to people with a certain condition—depression, diabetes, sleep deprivation—as performing lower than people without that condition on some cognitive tests; research papers might sometimes phrase it as people with that condition showing *deficits* on cognitive

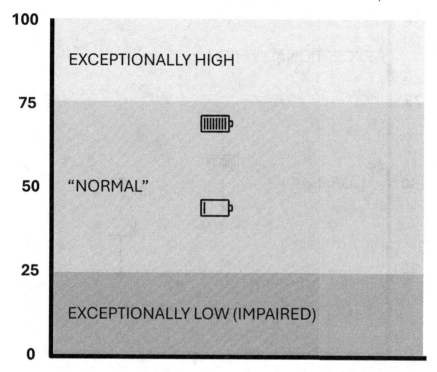

Figure 3.1 Hypothetical decline in cognitive performance when an individual is healthy vs. in a cognitively depleted state

functions. Again, just because the group with the condition performs, on average, lower than the group without the condition, this does not mean that the group with the condition was impaired. For example, on the same memory test as before, shown in Figure 3.2, perhaps the group without the condition has an average score of 55, shown by the fully charged battery symbol, while the group with the condition has an average score of 35, shown by the low-battery symbol. The average for the group with the condition, while lower than the average for the group without the condition, is within the normal range, and is not impaired. Also notice that the performance of people within each group spans a range, as shown by the arrows, and that very few people in the group with the condition actually do fall in the impaired range. In fact, some people with the condition perform as well as, and even better than, some of the people without the condition. It is the *average* performances of the two groups that differ.

To summarize, the cognitive changes we experience from conditions that affect our physical and psychological functioning are real and can be very disruptive, but they do not necessarily mean that we are impaired. The Appendix will explain how a neuropsychological evaluation addresses these questions.

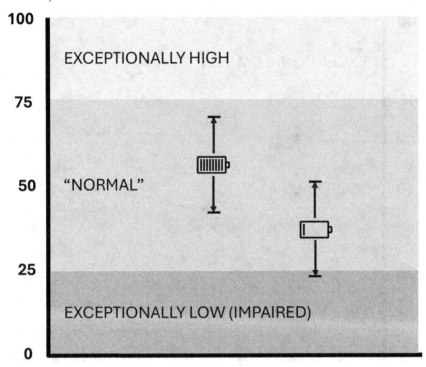

Figure 3.2 Hypothetical differences in cognitive performance between a healthy
group and a group in a cognitively depleted state

A Demanding Life

A second reason our healthy brains glitch is because many of us are living
lives that impose unreasonable, excessive demands that exceed even a
healthy brain's capacity.

As we reviewed in Chapter 1, our brains evolved to keep us alive by
anticipating threats to our survival and well-being in the environment, and
by developing flexible adaptations to keep us alive and well. We evolved in
a context where challenges to survival were immediate and concrete—predators, water, food, shelter. We are now operating those brains in a world
that floods us with virtual and socially constructed "threats." From global
tragedies to the personal (though carefully curated) lives of thousands of
people, we hold it all literally on the palm of our hands. Notifications and
breaking news on the screens we are constantly staring at hijack our brain's
alarm systems, which evolved to monitor the environment for signals of
trouble in order to protect us. We are constantly inhibiting distractions,
constantly shifting our attention. Work reaches us at home, the world
reaches us everywhere, we are rarely, if ever, unavailable or disconnected
from an influx of information from anywhere around the globe—and yet

many of us lack close social supports and are handling our multiple roles and responsibilities more alone than ever, all with a pervasive sense of urgency and rushing.

We glitch because we are using our brains in ways they were not designed for. The most expert driver, driving the most sophisticated, fine-tuned automobile, cannot stop it from stalling if they drive it through deep enough water.

* * *

The result of these two forces is a problem of decreased capacity and excessive demands, a depleted brain struggling with the cognitive overload of a demanding world. It is like we are running too many apps in our brain—the stress app, the depression app, the chronic pain app, the sleep deprivation app, the another-school-shooting app, the not-enough-savings app, the nemesis-from-high-school-looking-great-on-Instagram app, the aging-parents app. And just like our phones or computers when we run too many processes at the same time, we slow down, we freeze, we glitch.

The Two Pillars of a Brain-Friendly Life

Implementing changes so we can live a brain-friendly life might seem challenging, but think about it this way: Your brain is like a safe box where you keep everything that is most precious to you—the memories that make you who you are, your feelings for the most important people in your life, your ability to appreciate art that moves you to tears, your capacity to think, speak, relate, create, and love. Isn't the device that makes your most treasured experiences possible worth treating with utmost care?

Tending to Our Depleted Brains

In a survey of the most common recommendations clinical neuropsychologists make to patients, the most frequent recommendation was to engage in activities known to improve mood, followed closely by adherence to medications, both recommended often or always by over 80 percent of neuropsychologists.[12] Similarly, guidelines for cognitive rehabilitation for patients recovering from traumatic brain injury start by addressing factors that impact cognition, like hearing and vision impairments, fatigue, sleep disturbance, anxiety, depression, pain, and substance and medication use.[13]

Why is this? Because managing our overall health is crucial for our cognitive functioning. The first pillar of a brain-friendly life is to unburden our brain by tending to the body it lives in and the mind that it is inextricably linked to.

Managing conditions that drain the brain's resources is particularly important when we are trying to figure out if a person's cognitive changes could indeed be due to brain disease. One common—and frustrating—situation I encounter as a neuropsychologist is when I complete a neuropsychological evaluation and I have to conclude that the person has to be re-evaluated once their depression, anxiety, post-traumatic stress disorder, substance use, insomnia, and chronic pain are well controlled, when they are not taking multiple doses of opioids a day, or when they are not smoking marijuana from morning to evening. These factors, often referred to as *modifiable factors*, can affect how people perform on cognitive tests and obscure their actual cognitive capacities, mimicking or masking the symptoms of a possible brain disease. Cognitive testing is like a stress test for the brain: You would not show up to a stress test after spraining your ankle, because it would affect how you perform. Similarly, we cannot determine what your brain can and cannot do if it is bogged down by these conditions.

While these factors have immediate effects on our daily functioning, there is another important reason to address them: Many of these conditions, especially if untreated or poorly managed, increase our risk for cognitive decline and the development of dementia over time. In fact, it has been estimated that addressing modifiable risk factors like excessive alcohol consumption, depression, and chronic health conditions could prevent or delay a third or more of all dementias.[14]

Reducing Cognitive Overload

The strategies to tend to our brain by addressing modifiable factors are long-term strategies, ways of living and caring for our bodies and minds. The main focus of this book, however, is how to make our days more brain-friendly by implementing strategies that can reduce cognitive overload *today*, independently from our ongoing efforts to tend to our health. These strategies can help whether we are healthy and simply dealing with cognitive overload, or whether we are also experiencing the effects of chronic sleep deprivation, chronic stress, aging, etc.

Healthy Concern vs. When Worry Takes Over

None of the information presented in this book is meant to override any concerns you might be having about your cognitive functioning. Insight is a good thing. Remember that self-awareness and self-monitoring rely on executive functions, so it is a good sign to be concerned—it means your self-monitoring systems are healthy and active. As we mentioned in Chapter 2, many individuals with cognitive dysfunction due to brain disease have little or no insight into their impairments.

Paying attention to cognitive concerns is also important because in some cases they indeed are a very early sign of a neurocognitive disorder that reveals itself years later. The term *subjective cognitive decline* is used to refer to people's perception of their cognitive functioning as declining, even though they perform within normal limits on cognitive tests. There is no "objective" evidence of cognitive decline, only their subjective report.[15] Most individuals reporting subjective cognitive decline do not actually decline into mild cognitive impairment or dementia over time, but some of them do. So talking to your physician about your concerns and making an informed decision about next steps is the way to go (see the Appendix for more information about this process). You can work on addressing the modifiable factors in Part II and implementing the strategies in Part III, regardless of where you are in this process, whether you are not quite ready to seek an evaluation, whether you are waiting for an already scheduled evaluation (unfortunately long waits are common), or whether you already had an evaluation and were told there is no concern for a neurocognitive disorder.

However, two things are true: It is important to acknowledge our concerns and focusing too much on our cognitive functioning can make things worse. Once we start noticing our glitching and worrying about it, our anxiety can make us glitch more and can distort our self-perceptions. During a neuropsychological evaluation, for example, it is not uncommon for patients who are quite anxious about their cognitive performance to think they are "failing" a test, to the point of tearing up or needing to take a break, when in fact their performance is completely within normal limits.

Health professionals are familiar with this experience. You might have heard the term "worried well" (although the term is appropriately falling into disuse) to refer to people, often high-functioning individuals, who experience significant and long-standing concern that something is medically wrong with them—in this case, that something is wrong with their brains. [16] In some cases, this concern can rise to the level of *cognitive hypochondriasis*, when concerns are resistant to repeated reassurances that they are healthy, and every little mistake and lapse appears to confirm they have a serious neurological problem.[17] Some people are at higher risk of developing this kind of persisting concern—among others, those with histories of depression, anxiety with panic attacks, and childhood physical and sexual abuse; those experiencing high levels of stress; and those with low levels of social support.

A related but distinct experience is that of patients with so-called *functional (neuro)cognitive disorders*.[18] The diagnosis of functional neurological symptom disorder, or simply *functional neurological disorder*, applies to patients who display neurological symptoms (like paralysis, tremors, or seizures) in the absence of a neurological explanation, sometimes in a manner actually inconsistent with known neurological functioning or disease—their symptoms do not quite make sense, from a neurological perspective.[19] These patients are not faking their symptoms—their experiences are genuine and their symptoms real—but they are not caused by a neurological condition. The term *functional (neuro)cognitive disorder* has been proposed, by extension, to refer to patients who present with cognitive complaints and sometimes profound cognitive impairment on testing, in the absence of a neurological or other medical explanation.

In all of these cases—whether a person is experiencing mild concern about cognitive lapses, subjective cognitive decline, cognitive hypochondriasis, or functional cognitive disorders—a reasonable and helpful first step is obtaining information: Information about brain function, brain disease, factors and conditions that affect brain function, and the strategies that can help relieve a depleted brain. You have taken that first step, so let's continue.

Resources

- The website for the McKnight Brain Research Foundation (www. mcknightbrain.org) has helpful information about brain health, and long-term strategies to preserve it. While the focus is on aging, the advice is helpful for any life stage.
- *The Distracted Mind: Ancient Brains in a High-Tech World*, by Adam Gazzaley and Larry D. Rosen, is a thorough exploration of the many ways our modern world is a bad fit for our brains.
- Similarly, *Attention Span: Find Focus, Fight Distraction*, by Gloria Mark explains the nature of our attention and its executive control, and delves into the tension between our brains and technology, including the myths, limitations, and costs of constant multitasking.
- If you have been diagnosed with functional cognitive disorder or another functional neurological disorder, the website www.neurosymp toms.org, by neurologist Jon Stone, has helpful information regarding these disorders and how to manage them.

References

1. Elmo [@elmo]. (2024, January 29). Elmo is just checking in! How is everybody doing? [Post]. *X*.
2. O'Kane, C. (2024, January 31). Elmo wrote a simple tweet that revealed widespread existential dread. Now, the president has weighed in. *CBS News*. www. cbsnews.com/news/elmo-tweet-replies-how-is-everybody-doing-existential-dread-president-joe-biden-response.
3. Colosi, R. (2024, January 30). Elmo's viral tweet sparks an existential crisis among his followers. *Today*. www.today.com/parents/celebrity/elmos-viral-tweet-sparks-e xistential-crisis-rcna136346.
4. Fish, R. (2024, January 30). The internet is trauma-dumping on Elmo. *The Hollywood Reporter*. www.hollywoodreporter.com/news/general-news/elmo-how-a re-you-doing-twitter-1235811025.
5. parker (Taylor's version) [@pmilbs_]. (2018, March 15). One of my favorite games to play is "is my headache from dehydration, caffeine withdrawal, lack of proper nutrition, my [Post]. *X*.
6. trash jones [@jzux]. (2021, April 24). wtf is a "group chat?" i am an adult, I only have two friends and they don't know each other [Post]. *X*.

7. chuuch [@ch00ch]. (2019, January 27). sorry can't come out tonight I gotta sit alone in front of a muted tv opening and closing the same [Post]. *X*.

8. aubrey [@aubreybell]. (2019, December 5). oh boy ever spill a little bit of your coffee and realize the thread you are hanging on by is [Post]. *X*.

9. Voormolen, D.C., Cnossen, M., Polinder, S., Gravesteijn, B.Y., Von Steinbuechel, N., Real, R.G.L., & Haagsma, J.A. (2019). Prevalence of post-concussion-like symptoms in the general population in Italy, The Netherlands, and the United Kingdom. *Brain Injury*, 33(8), 1078–1086.

10. Asken, B.M., Snyder, A.R., Clugston, J.R., Gaynor, L.S., Sullan, M.J., & Bauer, R. M. (2017). Concussion-like symptom reporting in non-concussed collegiate athletes. *Archives of Clinical Neuropsychology*, 32, 963–971.

11. Willigenburg, N.W., & Poolman, R.W. (2023). The difference between statistical significance and clinical relevance. The case of minimal important change, non-inferiority trials, and smallest worthwhile effect. *Injury*, 54, Article 110764.

12. Meth, M.Z., Bernstein, J.P.K., Calamia, M., & Tranel, D. (2019). What types of recommendations are we giving patients? A survey of clinical neuropsychologists. *The Clinical Neuropsychologist*, 33(1), 57–74.

13. Ponsford, J., Velikonja, D., Hanzen, S., Harnett, A., McIntyre, A., Wiseman-Hakes, C., Togher, L., Teasell, R., Kua, A., Patsakos, E., Welch-West, P., & Bayley, M.T. (2023). INCOG 2.0 guidelines for cognitive rehabilitation following traumatic brain injury, part II: Attention and information processing speed. *Journal of Head Trauma Rehabilitation*, 38(1), 38–51.

14. Livingston, G., Huntley, J., Sommerlad, A., Ames, D., Ballard, C., Banerjee, S., Brayne, C., Burns, A., Cohen-Mansfield, J., Cooper, C., Costafreda, S.G., Dias, A., Fox, N., Gitlin, L.N., Howard, R., Kales, H.C., Kivimäki, M., Larson, E.B., Ogunniyi, A., … Mukadam, N. (2020). Dementia prevention, intervention, and care: 2020 report of the *Lancet* Commission. *Lancet*, 396, 413–446.

15. Jessen, F., Amariglio, R.E., Buckley, R.F., van der Flier, W.M., Han, Y., Molinuevo, J.L., Rabin, L., Rentz, D.M., Rodriguez-Gomez, O., Saykin, A.J., Sikkes, S. A.M., Smart, C.M., Wolfsgruber, S., & Wagner, M. (2020). The characterisation of subjective cognitive decline. *Lancet Neurology*, 19(3), 271–278.

16. Gray, D.P., Dineen, M., & Sidaway-Lee, K. (2020). The worried well. *British Journal of General Practice*, 70(691), 84–85.

17. Boone, K.B. (2009). Fixed belief in cognitive dysfunction despite normal neuropsychological scores: Neurocognitive hypochondriasis? *The Clinical Neuropsychologist*, 23(6), 1016–1036.

18. Ball, H.A., McWhirter, L., Ballard, C., Bhome, R., Blackburn, D.J., Edwards, M.J., et al. (2020). Functional cognitive disorder: Dementia's blind spot. *Brain*, 143, 2895–2903.

19. American Psychiatric Association. (2022). Somatic symptom and related disorders. In *Diagnostic and statistical manual of mental disorders* (5th ed., Text Revision). American Psychiatric Association Publishing.

Part II

Tending to Our Depleted Brains

Over the last couple of decades, we have done a great job of increasing awareness and public education about neurological conditions like dementia and concussions. But we have not always similarly emphasized what "normal" healthy brains look like, the effects of non-pathological processes like aging and stress on the brain, and the effects of health conditions on our cognitive functioning. The next part of this book addresses that. The issues covered in the following chapters share in common that they can drain our brain's resources, reducing its capacity to efficiently carry out its most complex functions. The first component of building a brain-friendly life is to unburden our brain by addressing these conditions.

In her book *Kitchen Table Wisdom*, Dr. Naomi Remen says, "No gardener ever made a rose. When its needs are met, a rosebush will make roses. Gardeners collaborate and provide conditions which favor this outcome".[1] This is a perfect way to describe what this next part of the book is about. You cannot really make a brain-friendly life out of nothing, any more than a gardener can make a rose out of nothing. There is no quick magic formula to develop a healthier brain, any more than there is a quick formula to grow a healthy rosebush. The best we can do is tend to the garden, creating the conditions that give our brain the best chance at a healthy life. The information and advice provided in the chapters in this section are meant to help you do just that.

Reference

1. Remen, R.N. (2006). Kitchen table wisdom: Stories that heal. Riverhead Books.

DOI: 10.4324/9781003409311-6

4 Chronic Stress

One of the many very relatable comics in the *Heart and Brain* series by Nick Seluk depicts Heart effortfully carrying a growing pile of bags labeled "past trauma," "stress," and "bad news."[1] When he approaches a small step labeled "minor inconvenience," Heart falls to the ground, sobbing. Brain finds him there and comments, "I think you're overreacting."

Chronic stress can act like that, draining our resources to the point that it renders us unable to cope with the seemingly minor inconveniences inevitable in daily life. When our cognitive resources have been drained, we might be unable to cope with situations that we would be able to handle under different circumstances, when we are not depleted. Because of this, chronic stress can be a major cause of cognitive lapses in daily life.

* * *

Every year, the American Psychological Association (APA) conducts a *Stress in America* survey. The most recent findings[2], [3] show that, on a scale where 1 is *Little to no stress* and 10 is *A great deal of stress*, most adults rate their average stress as a 5. However, one in four rate their average stress between an 8 and a 10. Parents and single adults are more likely to report high levels of stress. Women routinely report higher levels of stress than men, but the gender differences tend to be small; for example, 27 percent of women report high levels of stress compared to 21 percent of men.

There is an interesting age trend, with rates of high stress decreasing with age: One in three adults under 45, one in five adults ages 45 to 64, and only one in ten of those 65 and older describe their stress as an 8 or higher. Especially among adults under the age of 45, negative effects of stress are very common, including difficulty focusing, having less patience for others, feeling completely overwhelmed, feeling numb, being unable to do anything, or feeling so stressed they cannot function.

While financial and health-related problems are consistently among the most common stressors, more than 80 percent of adults identify global issues as another source of significant stress. Two-thirds feel their life has

DOI: 10.4324/9781003409311-7

been forever changed by the Covid pandemic and almost 90 percent feel there has been "a constant stream of crises" in the last couple of years.

We also seem to have a tendency to downplay stress: About two-thirds of adults feel their problems are not "bad enough" to be stressed because others have it worse. Over 60 percent do not talk about their stress to avoid burdening others, and as many say people around them just expect them to get over their stress. A third do not know where to start when it comes to managing their stress, and also a third feel completely stressed out no matter what they do to manage their stress. Perhaps unsurprisingly, over 40 percent acknowledge that they rely on unhealthy habits, like drinking more alcohol, to cope with stress.

These findings led the APA to conclude that we are "emotionally over-whelmed" and experiencing the psychological impacts of collective trauma related to the Covid pandemic, global conflict, racial injustice, inflation, and climate-related disasters.[2] What does this mean for your brain?

What Stress Is, and What It Was Not Meant to Be

Stress refers to the experience of being in—or anticipating—a situation where the demands exceed our resources. The *stressor* refers to the event or situation itself that causes stress.[4] Notice a couple of things. First, the same situation can be stressful or not depending on your resources at that time: If we can easily handle a situation, it will not be stressful, but if we feel we don't have the physical or psychological resources to handle the situation, it will be experienced as stressful. For example, our child coming home from school and announcing they forgot they have a project due the next day can be experienced as anything from mildly annoying to highly stressful, depending on whether we are healthy, rested, and have nothing else scheduled that afternoon, or whether we only got five hours of sleep the night before, we feel a migraine coming, and we offered to run errands and babysit for our neighbor, who is recovering from a car accident.

Second, even positive and welcome events can be stressful: A promotion, a new baby, a new puppy, planning a big family celebration, or a good friend unexpectedly passing through town on a weekday and wanting to see us can all be experienced as stressful if they strain our limited resources available. The key issue in determining whether and to what extent some-thing is stressful is "Can I do this? Can I handle this?"

Because stress is inevitable, our bodies evolved a way to handle it. The *stress response* has psychological and physiological components.[5] The psychological components include our subjective emotional experience of stress: Do we feel tense? Panicky? Helpless? Excited? It also includes a cog-nitive component—the thoughts we have about the situation. Are we telling ourselves, "This is it, I can't do this anymore"? Or "Why is this on me? Why does *nobody ever* help me?" Or "All right, this is going to be tough, but I've got this"? Finally, there are behavioral components to the psychological response, the modern versions of fight, flight, or freeze. What do we do? Do we roll up our sleeves and tackle the problem head on? Do we pretend it is not happening, "check out," and procrastinate? Do we become paralyzed, psychologically by becoming unable to start the simplest chore, or physically by developing debilitating stress-related symptoms?

The physiological aspects of the stress response involve a cascade of changes involving sympathetic, endocrine, cardiovascular, and immunological systems.[5], [6], [7] Hormones including cortisol, epinephrine (adrenaline), and norepinephrine are released; heart rate, cardiac output, blood pressure, and blood glucose levels increase; blood flow is directed to large skeletal muscles; digestion is slowed; the immune system is activated. All of this happens in preparation for action—fighting or fleeing as needed—and in anticipation of possible injury. In the brain, there is activation in limbic areas that orient our attention towards salient, emotionally relevant information important for our survival and well-being.

As you are probably well aware, there are individual differences in *stress reactivity*, the magnitude and time course of our body's physiological response to stress. Some of us show more intense neural and cardiovascular reactions to stress, for example, and take longer to recover, meaning these physiological responses persist longer after the stressor is gone. Interestingly, both ends of the spectrum are associated with adverse health consequences: Those with high physiological stress reactivity might be at higher risk for diseases like hypertension, but blunted physiological reactivity has been associated with behavioral outcomes like depression and substance dependence.[5]

This means stress responses of a certain, moderate intensity are natural and healthy. Our complex stress response, while often experienced as unpleasant, is an adaptive feature of our brain, meant to keep us alive and well. An organism that does not feel stress would not survive.

Chronic Stress

Our stress response mechanisms, adaptive and healthy in the short term, can trigger pathological changes when they become chronically active. Prolonged stress responses can result in cumulative strain on our cardiovascular, endocrine, immune, and other body systems, and in dysregulation of the neurobiological stress response itself. The term *allostatic load* is sometimes used to refer to this wear and tear on the body.[5], [6] Because so many body systems are involved, chronic stress increases our risk for a broad range of conditions including depression, heart disease, stroke, immune problems, and digestive issues, and can exacerbate almost any preexisting condition.[2] Chronic stress can also make us more sensitive to daily hassles and change our life outlook by making us focus solely on the day-to-day, losing sight of our long-term goals (what we sometimes call, colloquially, "survival mode").

Chronic stress does not affect just us: It affects those around us. Humans experience *emotional contagion*, the automatic and involuntary transmission of emotional states between individuals.[8] When we observe someone who is stressed, we experience some of the same stress-related physiological responses they do. Even infants, when their mothers hold them while experiencing the physiological stress reaction, show an increase in heart rate.[6] This has been referred to as *empathic stress* or *second-hand stress*, and the effect is stronger when we are emotionally close to the stressed person and for those of us with higher levels of empathy.

So, clearly, chronic stress is harmful to our mental well-being, physical health, and relationships, correct?

Not so fast.

Our *beliefs* about stress make a difference in how harmful stress is. A study examining the relationship between stress levels and mortality over the next eight years found that not everyone who experienced high stress levels was more likely to die prematurely: It was people who experienced high stress levels *and* who believed stress was harmful for their health that had a 43 percent higher risk of death.[9]

Stress mindset refers to our beliefs about the nature of stress, specifically, whether we believe that stress can enhance our functioning—for example, through opportunities to learn and become resilient—or whether we see stress as purely debilitating. People with more positive views of stress show less distress when experiencing adverse life events, perform better academically and cognitively when under stress, experience lower levels of anxiety and depression, report lower overall levels of perceived stress, and show a healthier pattern of physiological stress response and faster recovery.[4], [10]

Why would stress mindset impact psychological and physical health to this extent? The mechanisms are complex, but one avenue seems to be that when we embrace positive stress beliefs, we are more likely to engage in so-called *approach coping*, meaning when faced with a stressor, we respond in an active, problem-solving way. This increases our feelings of *self-efficacy*, the belief that we have the resources to handle a stressor (the feeling of "I can do this"). Those of us with this kind of mindset might also have a more active approach to our own health—for example, we are more likely to be consistently physically active, and we tend to experience more positive mood, which is known to have positive effects on health.

The good news is that stress mindsets can change. In experimental studies, just briefly educating participants on the positive aspects of stress—for example, that its physical sensations are a healthy body's way to ready us for action and promote our well-being—can result in healthier responses during a stressful task. The key message is that while chronic stress can be draining, there are things we can do to minimize its psychological and physical effects.

Let me pause here and acknowledge something. As I write this, what comes to mind is a meme that shows the tweet:

> Stop saying "I wish" and start saying "I will"[11]

followed by the reply:

> I will my parents still loved me.

I understand that for those of us enduring difficult and sometimes painful chronic stressors outside of our control, talk of changing our stress mindset can sound like being told to "just think positively." Are we supposed to write positive thoughts in our gratitude journal about losing our job, or about mom's cancer treatment? No. In Chapter 18, we will address what changing mindsets and beliefs actually looks like. For now, let me say this: It might be difficult or impossible to think differently about the stressors, but you can think differently about your stress response: Your stress response is normal. The beating heart, the sweaty palms, the flushed face are not signs that there is something wrong with you. That is the feeling of your brain trying to keep you alive, safe, and well.

How Stress Affects Cognitive Functioning

During the acute stress response, cortisol can cross the blood–brain barrier and bind to receptors in areas crucial for executive function, memory, and emotions.[12] The impact of acute stress on cognition is complex, but as a first principle, either too high or too low levels of stress hormones can depress performance. An optimal level of stress is required for an optimal level of cognitive functioning.

Generally, acute stress activates the *salience network*. This is an attentional network that orients our attention towards novel stimuli in the environment. This makes sense: The brain needs a system that scans the environment for information that could be relevant to our survival and well-being, like a sudden noise or movement out of the corner of our eye.[13] Because of this, cognitive functions like sensory attention and vigilance are actually enhanced. Also activated are neural systems important for rapid, automatic behavioral responses.

In contrast, stress suppresses activity in prefrontal areas needed for many executive functions, including working memory, cognitive flexibility, and complex and creative problem-solving.[14] The general picture is that under acute stress we become cognitively inflexible and act based on habit.[15] The brain prioritizes rapid but rigid action. If you come home at the end of the day in a state of heightened stress, you are less likely to think thoughtfully about making a well-balanced meal for yourself that includes multiple ingredients. You might instead just pick up a drive-thru meal—the same thing you always get, because you don't have the bandwidth to scan the menu and try something different.

The effect of stress on memory is interesting. In general, acute stress seems to enhance *encoding* and *consolidation*—that is, forming new memories and "saving" them in long-term memory, especially for information related to the stressor.[14], [15] Again, this makes sense: Stress is a signal that something important and potentially threatening is happening, and our brain wants to remember the situation so it can be prepared if we encounter it again in the future. In contrast, *retrieval* is suppressed.[7] This is why—as anybody who has ever taken an important test knows—when we are stressed, it is difficult to recall information. Stress doesn't make us forget things we know, but it makes it more difficult to recall or retrieve that information. It can also make it more difficult to retrieve words, particularly under time pressure, and those frustrating "tip of the tongue" lapses are common when we are stressed, especially as we get older.[12]

These acute cognitive effects, while perhaps annoying, are normal and adaptive. In healthy, short-lived stress responses, as the level of stress hormones decrease, cognitive functions return to normal. But in individuals with chronically elevated stress levels, ongoing cognitive complaints are common. They show decreased performance in executive functions including working memory, cognitive inhibition, and attention shifting, and they experience more attention lapses.[13], [16] Their attention is also more sensitive to negative stimuli—their brain is scanning the environment for signs of trouble, which distracts from whatever they are trying to do in that moment.

Unsurprisingly, changes in the brain itself have been documented in people under chronic stress, including decreased volume and connections within gray matter, especially in prefrontal areas, and changes in white matter integrity; in contrast, there is increased volume and connections in areas responsible for automated and habitual behaviors.[7], [12], [13], [16] Chronic stress also inhibits *neurogenesis*, the formation of new neurons.[14] In older adults, persistently high stress is not just associated with lower cognitive performance, but faster cognitive decline over time.[17] Luckily for us, our brain's *plasticity*—its ability to change in response to experience—works both ways, and brains can be rehabilitated through positive changes and by practicing healthy responses to stress.

What To Do About It

In Part III, we will cover strategies that can help reduce glitching due to stress by easing the cognitive demands on our brain, freeing up cognitive resources, managing the acute physiological and psychological stress response, reframing specific stressors in healthier ways, and building a healthier stress mindset. In addition, here are a few things you can do in the longer term, to address the more chronic aspects of stress, develop a healthy approach to stress management, and prevent harmful effects of stress.

1. Assess your stress. Many of us underestimate the magnitude of the stress we experience. Ask yourself questions like, how often do you feel …

• …you are unable to control important things in your life?

- ...you cannot handle your problems?
- ...things are not going your way?
- ...you cannot keep up with things, that is it all just too much?[18]

Other important aspects of your stress experience worth exploring are:

- How long have these stressors been going on? Are they temporary or chronic? Is there an end in sight?
- What is under your control? What is not?
- Who else is experiencing this? Are you carrying this stress alone?
- What resources do you have to address this? Think broadly: Do you know what to do? Do you know where to find information you need? Are you rested? Are you sick? Are you struggling with your mood? How much time do you have to devote to these problems? Do you have people who provide emotional support? Do you have people who provide practical help?

Based on your answer to these questions, do you feel, overall, you have the resources—psychological and practical—to deal with the stressors on your plate?

Next, examine your *coping style*. What is your natural, automatic response when experiencing stress? Some coping strategies focus on the problem or stressor itself—for example, coming up with a plan to deal with the situation more efficiently. Some strategies focus on our response—for example, by doing breathing exercises to calm down, working out to release tension, journaling, or venting to a friend. And some strategies focus on neither—for example by avoiding the stressor through distraction and pro-crastination, or avoiding our feelings by numbing with substances. Some ways of coping tend to be healthier and lead to better outcomes. In general, more active strategies, like problem-solving and reframing stressors (in the ways we will discuss in Chapter 18) tend to be more helpful and healthier than avoidant strategies.

Imagine, for example, you or a loved one are newly diagnosed with a health condition requiring surgery. Do you tend to...

- ...jump into action? Do you roll up your sleeves, research everything there is to know about the condition, treatment options, and outcomes, find a specialist, and reach out to friends to come up with a support plan for the recovery period?
- ...pretend it is not happening? Is your automatic response to say, "It's fine," then avoid talking about it? Do you escape by scrolling endlessly on your phone or overworking so you don't have a minute to think?
- ...numb with substances like alcohol or sedatives?
- ...do something else? If so, what is it?

As you go through the strategies in Part III, try to identify the ones that are the best fit for what you need. If your tendency is to avoid and procrastinate, pay special attention to strategies that help you plan and problem-solve instead. If your tendency is to overthink and ruminate, pay attention to strategies for the healthy reframing of stressful situations. If you are a "doer" but tend to ignore the stress on your body, pay attention to ways to ground yourself in your body and senses. The goal is to develop a varied and flexible repertoire of stress management strategies. As we will see, sometimes even avoidance is helpful, when deployed in the short term as needed before switching to a problem-focused strategy.

2. Consider seeking professional help. While this book has separate chapters for stress and mental health, stress and stress management are important components of our mental health. Depending on our experience and personal context, we might benefit from professional help. A counselor or therapist can help us, among other things, to:

- examine our stress responses, especially the emotional and cognitive components;
- examine our coping tendencies and how they help or hinder;
- learn new coping strategies to develop a broader and more flexible stress management style;
- identify untapped psychological strengths and resources to cope with stress;
- challenge and reframe unhelpful beliefs about stress; and
- explore and address how our stress and attempts at managing it might be impacting our health.

See Chapter 5 for guidance on how to find a counselor or therapist.

3. Tend to the garden. Adequate sleep and physical exercise, in particular, improve our ability to regulate our emotions, thoughts, and physical reactions. Physical exercise improves brain connectivity and functioning, and can be thought of as a powerful antidote to the noxious effects of chronic stress. In addition, we are programmed to turn to others in times of stress or when experiencing stress-related depletion. There is evidence that after we have a stressful experience, our attention to signals of social threat decreases and social aspects of cognition are enhanced: We get better at recognizing other people's emotions, we are more likely to detect happy facial expressions, we are more empathic, and we are more likely to trust others.[7] This has been referred to as the "tend-and-befriend" aspect of the stress response.[19] Seeking support and connection is what we are meant to do.

Social support can be particularly helpful for stress due to specific situations, like parenting a child with special needs, living with a chronic illness, divorce, bereavement, single parenthood, and caregiving for someone with a terminal illness, serious mental illness, or dementia. Connecting to others in the same situation—for example, through support groups—does more than provide emotional support: We can learn from each other how to manage difficult situations, increasing our self-efficacy, which leads to more active ways of coping and ultimately results in a healthier stress experience.

Resources

- To get you started thinking about longer-term brain maintenance and "tending to the garden," *The Brain Health Book: Using the Power of Neuroscience to Improve your Life*, by John Randolph, is a good resource to learn more about lifestyle factors—like exercise, mentally stimulating activities, and nutrition—that can preserve brain health and prevent cognitive decline. See also the books listed under *Resources* in Chapter 9.
- The American Psychological Association's website (www.apa.org) has extensive information on stress and stress management, including specific resources like video guides for management of acute stress. The yearly *Stress in America* surveys can also be found on their site.
- The American Heart Association's website (www.heart.org) also has helpful information and resources regarding stress, health, and stress management.

References

1. Seluk, N. (2022, July 26). The Build Up. *The Awkward Yeti*. https://theawkwardyeti.com/comic/the-build-up.
2. American Psychological Association. (2023). Stress in America 2023: A nation recovering from collective trauma. www.apa.org/news/press/releases/stress/2023/collective-trauma-recovery.
3. American Psychological Association. (2022). Stress in America: Money, inflation, war pile on to nation stuck in COVID-19 Survival Mode. www.apa.org/news/press/releases/stress/2022/march-2022-survival-mode.
4. Jenkins, A., Weeks, M.S., & Hard, B.M. (2021). General and specific stress mindsets: Links with college student health and academic performance. *PLoS ONE*, 16 (9), e0256351.
5. Whittaker, A.C., Ginty, A., Hughes, B.M., Steptoe, A., & Lovallo, W.R. (2021). Cardiovascular stress reactivity and health: Recent questions and future directions. *Psychosomatic Medicine*, 83, 756–766.
6. Engert, V., Linz, R., & Grant, J.A. (2019). Embodied stress: The physiological resonance of psychosocial stress. *Psychoneuroendocrinology*, 105, 138–146.
7. Von Dawans, B., Strojny, J., & Domes, G. (2021). The effects of acute stress and stress hormones on social cognition and behavior: Current state of research and future directions. *Neuroscience and Biobehavioral Reviews*, 121, 75–88.
8. Dimitroff, S.J., Kardan, O., Necka, E.A., Decety, J., Berman, M.G., & Normal, G.J. (2017). Physiological dynamics of stress contagion. *Scientific Reports*, 7, 1–8.
9. Keller, A., Litzelman, K., Wisk, L.E., Maddox, T., Cheng, E.R., Creswell, P.D., & Witt, W.P. (2012). Does the perception that stress affects health matter? The association with health and mortality. *Health Psychology*, 31(5), 677–684.

10. Laferton, J.A.C., Fischer, S., Ebert, D.D., Stenzel, N.M., & Zimmermann, J. (2019). The effects of stress beliefs on daily affective stress responses. *Annals of Behavioral Medicine*, 54, 258–267.
11. Alvord, T. [@TiffanyAlvord]. (2016, June 30). Stop saying "I wish" and start saying "I will" [Post]. *X*.
12. Mikneviciute, G., Ballhausen, N., Rimmele, U., & Kliegel, M. (2022). Does older adults' cognition particularly suffer from stress? A systematic review of acute stress effects on cognition in older age. *Neuroscience and Biobehavioral Reviews*, 132, 583–602.
13. Girotti, M., Adler, S.M., Bulin, S.E., Fucich, E.A., Paredes, D., & Morilak, D.A. (2018). Prefrontal cortex executive processes affected by stress in health and disease. *Progress in Neuropsychopharmacology and Biological Psychiatry*, 85, 161–179.
14. Sun, M.-K., & Alkon, D.L. (2014). Stress: Perspectives on its impact on cognition and pharmacological treatment. *Behavioural Pharmacology*, 35, 410–424.
15. Arnsten, A.F.T. (2015). Stress weakens prefrontal networks: Molecular insults to higher cognition. *Nature Neuroscience*, 18(10), 1376–1385.
16. Echouffo-Tcheugui, J.B., Conner, S.C., Himali, J.J., Maillard, P., DeCarli, C.S., Beiser, A.S., Vasan, R.S., & Seshadri, S. (2018). Circulating cortisol and cognitive and structural brain measures. The Framingham Heart Study. *Neurology*, 91, e1961–e1970.
17. Kulshreshtha, A., Alonso, A., McClure, L.A., Hajjar, I., Manly, J.J., & Judd, S. (2023). Association of stress with cognitive function among older black and white US adults. *JAMA Network Open*, 6(3), Article e231860.
18. Cohen, S., Kamarck, T., & Mermelstein, R. (1983). A global measure of perceived stress. *Journal of Health and Social Behavior*, 24, 386–396.
19. Taylor, S.E., Klein, L.C., Lewis, B.P., Gruenewald, T.L., Gurung, R.A.R., & Updegraff, J.A. (2000). Biobehavioral responses to stress in females: Tend-and-befriend, not fight-or-flight. *Psychological Review*, 107(3), 411–429.

5 Mental Health

In one of the many poignant storylines in the hit show *Breaking Bad*, two planes collide mid-air due to an error by an air traffic controller, who had returned to work despite being grief-stricken by his daughter's accidental death. While it might be tempting to see this as an exaggerated device for dramatic effect, I have had patients with profound depression tear up when asked for the date or how many grandchildren they have, because they feel so mentally paralyzed that they cannot answer basic questions. What is the nature of the relationship between psychological conditions like depression and cognitive functioning?

* * *

Mental health conditions are common and one of the leading causes of disability worldwide.[1] Prior to the Covid pandemic, one in eight people in the world was living with a mental disorder, most commonly an anxiety or depressive disorder.[1], [2] During the pandemic, the number of people living with anxiety or depression rose significantly: As many as half of the world population was experiencing some form of psychological distress, a third endorsed stress, depression, and/or anxiety, and 1 in 4 reported symptoms of post-traumatic stress and sleep problems, although there was significant variability in the rates of these problems depending on socioeconomic variables.[3]

When it comes to mental health, however, it can be difficult to interpret global numbers, because the prevalence, experience, and very definitions of mental health and its disorders often vary across different cultures. In the U. S., the proportion of people living with a mental health condition varies significantly by age, from as many as half of adults under 45, to as few as 15 percent of adults over 65.[4], [5] (It is important not to infer that this is solely because older adults are less likely to experience psychological symptoms; the differences can also be due to other variables—for example, how likely people of different ages are to disclose psychological difficulties and seek help.) Like other parts of the world, during the pandemic, the U.S. saw an increase in mental health symptoms, sleep and eating disorders, and

DOI: 10.4324/9781003409311-8

emergency visits for mental health crises, drug overdoses, and suicide attempts.[6]

Despite the significant impact mental health conditions have on health and functioning, it is unfortunately very common for them to go untreated. Only half of women and 40 percent of men living with a psychological condition have received mental health services in the past year, with rates as low as 1 in 4 for some ethnoracial groups.[4] Even among those Americans with a serious mental illness like schizophrenia, fewer than two-thirds have received services in the last year.

Most mental health conditions can affect cognition in some way to different degrees. In this chapter, we will focus on the effects of relatively common conditions like depression, anxiety, and trauma-related disorders, but it is important to note that other conditions like bipolar disorder and schizophrenia can be associated with broader, more chronic, and more profound cognitive deficits requiring specialized evaluation and management.

What Is a "Mental Disorder"?

The American Psychiatric Association's *Diagnostic and Statistical Manual of Mental Disorders*—commonly referred to as the DSM—defines a "mental disorder" as "a syndrome characterized by a clinically significant disturbance in an individual's cognition, emotion regulation, or behavior" that is associated with "significant distress or disability in social, occupational, or other important activities."[7] In other words, the person is experiencing a cluster of symptoms involving patterns of thinking (like the hopeless and self-deprecating thoughts of a person with depression), feeling (such as the acute panic experienced by a person with PTSD during a flashback), and/or behaving (like the isolation and withdrawal of a person with social anxiety) that cause them emotional distress or interfere with their ability to work, study, maintain and enjoy relationships, or otherwise function in daily life.

An important component of the definition is that cultural norms have to be considered when evaluating someone for a possible mental diagnosis. For example, the father of one young patient I saw for the cardiothoracic transplant service confided to me that his son's sudden cardiac failure was due to witchcraft, and a patient I evaluated for the epilepsy program told me her seizures were punishment for a family tragedy she believed she was responsible for. People sometimes experience what could be considered auditory and visual hallucinations during religious rituals or acute bereavement. Cultural context is crucial to determine whether any of these experiences and beliefs are normal and commonly accepted in their culture or symptoms of a mental disorder.

To be formally diagnosed with a mental health condition, the person has to meet the *diagnostic criteria*. Diagnostic criteria include a list of symptoms characteristic of the disorder and other information, like how long the symptoms must have lasted and what other conditions or explanations need to be ruled out before making the diagnosis. For example, to be diagnosed with panic disorder, a person has to experience recurrent, unexpected panic attacks characterized by at least four symptoms out of a list of 13, including palpitations, sweating, shortness of breath, dizziness, fear of dying, etc.[8] The episodes should be accompanied by significant worry or changes in beha- vior—like avoiding public places—for at least a month, and other explanations, such as substance use or a cardiac condition, have to be ruled out.

However, psychological symptoms that do not meet diagnostic criteria, sometimes referred to as *subclinical,* can still be profoundly disruptive, and the person could still benefit from treatment. For example, a person whose panic episodes are characterized only by three symptoms—intense chest pain, severe shortness of breath, and an overwhelming fear of dying, for example—would benefit from behavioral and possibly pharmacological treatment, as indicated after careful evaluation.

Cognition in Depression, Anxiety, and Trauma

Cognitive symptoms are common in depression, to the point that problems thinking, concentrating, or making decisions are part of the diagnostic cri- teria for a major depressive disorder, along with other emotional and phy- sical symptoms like depressed mood, decreased enjoyment of previously pleasurable activities (*anhedonia*), sleep disturbance, fatigue, feelings of worthlessness, and others.[9] The most common findings are slow cognitive processing speed and broad executive dysfunction, including problems with attentional control.[10], [11] While there are other common cognitive complaints—in memory and problem-solving, for example—these are often due to slow processing speed and executive deficits. When cognitive pro- cessing speed slows down, people feel easily overwhelmed by information, and because of this, patients with depression often complain that they "can't think" or "can't concentrate."

Cognitive deficits tend to be worse in patients with more severe and more chronic depression, and in people who experience their first episode in midlife or older age. In older adults, cognitive deficits can be so severe that the term *pseudodementia* is sometimes used to refer to dementia-like cog- nitive impairment that is due to depression. At the same time, late-life depression is a known risk factor for Alzheimer's disease, so older adults with depression and cognitive changes should consider not only seeking treatment for depression but a neurological evaluation.

Anxiety disorders—phobias, social anxiety, generalized anxiety, panic disorder, and others—are the most common type of mental health disorder; almost 1 in 3 people will experience one in their lifetime.[12] These condi- tions are characterized by psychological symptoms of fear or anxiety about an imminent or anticipated threat, and physiological symptoms of auto- nomic arousal characteristic of the stress response. Behaviorally, anxiety disorders often lead to increasing avoidance of situations, which can significantly limit the person's social, educational, occupational, and recreational opportunities. There is substantial comorbidity between anxiety disorders and between anxiety and depression: As many as two-thirds of people with major depressive disorder also meet criteria for an anxiety disorder and vice versa.[12]

Anxiety disrupts the allocation of cognitive resources to a task, leaving us with less available attentional resources.[13] This makes sense: When the brain is busy running the "anxiety app" with its ticker tape of anxious thoughts, attention lapses and difficulty shifting are likely. For example, anxiety weakens *error monitoring*. Typically, milliseconds after we make a mistake, before we are conscious we made it, there is a particular electrical signal detected on EEG, but when people with anxiety are mildly stressed (waiting to receive feedback about how they did on a math test, for example), this signal is decreased, suggesting that they are so preoccupied that they do not detect the errors they are making.[14]

Trauma refers to experiences of actual or threatened death, serious injury, or sexual violence.[15] Most individuals who are exposed to traumatic events experience temporary symptoms—anxiety, emotional dysregulation, nightmares—that eventually resolve, but 10 to 20 percent develop persistent and debilitating symptoms.[16] Post-traumatic stress disorder (PTSD) is a complex diagnosis involving a wide range of possible symptoms including intrusive thoughts, memories, or dreams about the trauma; altered physiological arousal (the person might either feel "jumpy" or numb); negative thoughts about themselves or the world ("I'm all alone," "You can't trust anyone"); and a host of other negative emotional experiences (like anger and feelings of detachment from others).[15]

You might remember the concept of associative learning from Chapter 1. When an individual experiences a traumatic event, the intense physiological response experienced at the time of the trauma can become paired with previously neutral stimuli, which then act as conscious or unconscious triggers for a similar physiological response. Let's say we witness a horrifying accident in which a pedestrian is hit by a car. We might later experience a similar physiological response to the one we felt at the time we witnessed the accident, triggered when we walk by the same area, or when we see a man with the kind of moustache the victim had, or when we hear tires screeching. That area, the moustache, and the screeching sound were previously neutral stimuli that are now associated with the trauma, and now they can, by themselves, trigger the trauma response. Moreover, physical sensations themselves can become a trigger. The mild, normal physiological stress response we feel when our boss asks us a question in the middle of a meeting can now trigger a cascade of sensations that result in a panic attack. Eventually, simply anticipating a mildly stressful experience can cause anxiety and might lead to avoidance—for example, we start skipping meetings, social functions, or public spaces altogether.

People with PTSD often show decreased performance in attention, processing speed, memory, and executive tasks, and the severity of the deficits is often (but not always) related to the severity of the PTSD symptoms.[17], [18]

A special instance of trauma deserves mention: *Adverse childhood experiences* (ACEs) are events involving threat or deprivation, including physical or sexual abuse, neglect, or exposure to violence, for example.[19] ACEs are associated with increased risk for a myriad of negative health, psychosocial, and socioeconomic outcomes. We now know that exposure to chronic or traumatic stress during the neurodevelopmental period leads to a mis-programming of the stress system. In healthy developing brains, there is a built-in mechanism for the down-regulation of the stress response once danger has passed: Our stress response naturally rises and then wanes. But exposure to ACEs can cause the stress response to not shut off naturally. Because of this, adults with histories of ACEs are at increased risk of developing PTSD and other psychological and medical conditions in response to adverse events later in life.

The Depressed, Anxious, and Traumatized Brain

In Chapter 1, we mentioned that there are networks in the brain, including the limbic system, that detect and process salient, emotionally relevant stimuli in the environment, giving rise to the physiological, psychological, and behavioral aspects of the emotional response. For example, our emotional network detects a person walking towards us with a scowling face and clenched fists, which triggers a tight sensation in our stomach, a feeling of fear, thoughts like "I need to get out of here," and a behavior such as ducking into a store. We also mentioned that executive systems in prefrontal areas regulate these emotional networks, calming our bodies and minds when we realize we are safe.

A common feature of disorders like depression, anxiety, and PTSD is dysregulation in these brain systems.[20], [21], [22] It is a perfect emotional storm: The systems that detect salient stimuli in the environment are over-active, especially in response to negative stimuli (like a critical comment or losing in a game), while the systems that down-regulate those emotional networks are under-active and unable to short-circuit the negative spiral. When anxious or depressed, we show *mood-congruent biases*. We tend to direct our attention towards negative stimuli (like sad news) more than positive stimuli, we have difficulty disengaging our attention from negative stimuli to focus on a task, and we tend to interpret ambiguous stimuli (like an inexpressive face) as negative ("They don't like me").[22], [23], [24]

In depression, there is a parallel finding in *reward systems*. There is a network in our brain that detects potentially pleasurable stimuli and directs our behavior towards them (as when we open the games or social media apps on our phones) and an executive system that regulates reward-seeking behavior (making us close the apps and go back to work). Depressed individuals show hypoactivation in the reward-processing network and hyperactivation in the behavior-regulating network, which might explain why they do not experience pleasure from enjoyable activities, and why they do not seek enjoyable activities to begin with.[22]

Consistent with the idea that these conditions drain brain resources, there is evidence that our brains have to actually work harder to get things done. When performing cognitive tasks, depressed patients show hyperactivation in the *default mode network* of the brain, a network that is typically active when we are at rest. This means that when we are depressed, we have difficulty overriding our brain's "idle" setting to actively engage in a task—it is difficult for our brains to "get going." Moreover, when depressed patients perform well on cognitive tasks, there is often hyperactivation in prefrontal areas, suggesting that additional executive effort is needed to perform the task adequately.[11]

These neural patterns also suggest difficulty inhibiting *self-referential* processing, which is also mediated by the default mode network. Self-referential processing refers to a form of self-focus, a form of thought when we are either thinking about ourselves (our lives, our past experiences, our future plans) or trying to figure out how what is happening in the world relates to us, what it means for us. Excessive self-referential processing might be one process behind *rumination*—those perseverative, negative, unproductive thoughts like "It's all my fault" or "My life is ruined." When depressed or anxious, we often personalize things that have nothing to do with us, like a coworker's bad mood ("I did something wrong"). In PTSD, it can also be a sign of a brain preoccupied with our own safety.[12]

Finally, cognitive, emotional, and physical symptoms are also related to imbalances in a variety of neurotransmitters, including serotonin, norepinephrine, and dopamine.[20], [25] Serotonin is involved in the regulation of mood, cognition, and physiological functions, and its dysregulation can cause symptoms in all of these domains. You might recall from Chapter 4 that norepinephrine is a stress hormone that prepares us to respond to threats; when imbalanced, it can result in physical symptoms of anxiety, like palpitations and sweating. This is related to why people with anxiety often show indicators of negative physiological arousal, such as an exaggerated *startle response*—the way we jump if there is a loud noise while we are watching a horror movie. Norepinephrine is also involved in attention and mood modulation. Dopamine is important for executive functions like working memory and attentional control, and for the regulation of pleasure- and reward-seeking behaviors.

Remember that thanks to the brain's plasticity, these changes can be reversed. Medications, for example, target imbalances in neurotransmitters, and pharmacological treatment and psychotherapy have been shown to normalize volume, connectivity, and function in brain areas where abnormalities have been documented.[26] Just as adverse experiences can derail brain functioning, healing experiences can restore healthy neural settings.

* * *

As we have seen, depression, anxiety, and trauma can cause cognitive lapses through different mechanisms. There are biochemical changes in the functioning of the brain itself that cause breakdowns in attention control, executive functions, and emotional regulation, and that can cause slowing of overall processing speed. Rumination acts like an app running in the background, taking up precious mental space and resources. Paradoxically, our efforts to rein in our emotions can further deplete our limited cognitive resources. We spend considerable effort attempting to regulate our mood and behavior, trying to quiet our minds, to get ourselves going, trying not to cry, not to panic, not to remember the horrible thing that happened. In this sense, we are constantly multitasking, doing what we are doing while managing our emotional state. Cognition is also affected indirectly, due the poor sleep, fatigue, and physical aches that can accompany these disorders. But there are ways to reclaim our brains.

What To Do About It

Part III will cover strategies that we can use on a daily basis to reduce glitching by reducing the demands on our depressed, anxious, or trauma-preoccupied brains. Some of the strategies can also help us manage some of the symptoms associated with these conditions themselves, like the physical symptoms of acute anxiety or ruminative depressive thoughts. While these strategies can help us function better even if we are actively experiencing psychological symptoms, these conditions *can* be treated, and it is essential to do so. Some of us might experience a single episode of these conditions; for some of us, these will be recurrent or chronic conditions that we learn to manage over time. But help is available.

1. *Deciding to get help.* You might be wondering whether what you are experiencing is "really that bad." Here are a few thoughts:

First, remember the definition of a mental disorder:

- Are your emotions, thoughts, or behaviors upsetting to you?
- Are they interfering with your ability to interact with your loved ones, socialize, or work, or are you doing so with extreme effort?
- Are you not enjoying time with your loved ones (are you overwhelmed instead of delighted when your grandchildren visit)?
- Are you letting go of things that are important for you to be part of (like family occasions or your volunteering shift) to avoid anxiety, or because you just don't have it in you?

Second, our self-assessments—from how sleepy we are to how skilled of a driver we are—are not highly reliable in general, and might be even less so when we are experiencing psychological symptoms that affect the executive systems responsible for self-awareness. Moreover, many of us might have internalized stigma about mental disorders. Overall, we might not always be the best judges of how we are actually doing emotionally. (Once, while in graduate school, I went to see a very experienced psychiatrist for help with depression. After patiently listening to me for a while, he said, "It sounds to me like you're also feeling quite a bit of anxiety." Without missing a beat, I shook my head: "I'm not anxious, I just can't breathe." He looked me straight in the eye with a gentle smile on his face until I realized what I had said. Mind you, I was in the process of obtaining a Ph.D. in clinical psychology.)

Finally, psychological conditions can present differently in different people. For example, depression can present primarily with physical symptoms (especially in men and in older adults) or irritability (again, especially in men). PTSD can present as emotional hyper-reactivity or emotional numbness. And medical conditions like thyroid disorders and hormonal imbalances can mask, mimic, or exacerbate a psychological condition, making expert assessment necessary.

For all these reasons, if you are wondering if what you are feeling is "normal," seek consultation with a psychologist or another mental health professional, or bring it up with your primary care provider. They can help you examine what you are experiencing and advise you on next steps.

2. Getting help. Psychotherapy and psychopharmacology options are available. Depending on your symptoms, circumstances, and preferences, you can try psychotherapy alone first. If you are experiencing circumscribed physical symptoms of anxiety without major emotional difficulties, a short course of medication might be all that is needed. For most moderate to severe symptoms, however, a combination of medication and psychotherapy will usually be most effective.

Keep in mind that there is significant variability in outcomes, for both therapy and medication. If what you try first does not work, do not desist—there are options. Evidence-based psychotherapeutic approaches for different conditions include cognitive behavioral therapy, behavioral activation, mindfulness-based therapy, acceptance and commitment therapy, cognitive processing therapy, prolonged exposure therapy, eye movement desensitization and reprocessing, among others.

Medication options are similarly broad, and include antidepressant and antianxiety medications targeting serotonin and/or norepinephrine, other antidepressant classes, sedating benzodiazepines for anxiety and sleep disturbance, beta blockers for physical symptoms of anxiety, sleep aids including medication prescribed for PTSD-related nightmares, and many other options, all with different profiles of efficacy and side effects. Your primary care physician might be comfortable prescribing some of these medications; however, depending on the nature and severity of your symptoms, other medical conditions you have, other medications you are taking, and whether multiple medications might be needed, referral to a psychiatrist or an other prescribing professional might be preferred.

There are also new interventions—you might have heard about non-invasive brain stimulation and psychedelic therapy, among others. The decision-making regarding treatments is complex, and there is much misinformation, so make sure to turn to reliable sources (see below).

To find a mental health provider, speaking to your primary care physician is a good place to start, since they might have recommendations. Regional associations of mental health professionals—including psychologists, psychiatrists, counselors, and licensed clinical social workers—often have directories on their websites, where you can search for a provider based on characteristics like their location and their areas of expertise. Similarly, depending on your healthcare system, insurance plans might have directories of providers in your area. The American Psychological Association also has a psychologist locator that you can search by geographic area and specialty (see below).

3. Tend to the garden. Health behaviors like exercise, sleep hygiene, and healthy nutrition are included in guidelines for the treatment of depression. [27] Exercise in particular has been shown to improve symptoms of depression and anxiety, even in populations facing significant stressors, like cancer survivors.[28], [29] While exercise of higher intensity does seem to provide greater benefits, effects have been documented for many types of physical activity, including aerobic exercise, resistance training, and yoga. Finally, remember the importance of social connection. The availability of strong social support predicts better mental health and is thought of as a protective factor: It actually reduces our risk of experiencing a mental health condition, including our risk of developing trauma-related symptoms after a traumatic event.[30]

Resources

- Reliable sources of information on mental health and available treatments include the Substance Abuse and Mental Health Services Administration (www.samhsa.gov), the American Psychological Association (www.apa.org), and the American Psychiatric Association (www.psychiatry.org; see the section *Patients and Families*).
- The National Institute of Mental Health (www.nimh.nih.gov) has a section specifically on mental health medications under *Health Topics*.
- The American Psychological Association's Psychologist Locator is at www.locator.apa.org.
- The American Psychological Association also has a podcast, *Speaking of Psychology*, featuring interviews with psychologists who are experts on a wide range of topics. The podcast *Road to Resilience*, hosted by the

Icahn School of Medicine at Mount Sinai also has episodes on wide-ranging mental health topics.

- The books *The Other Side of Sadness: What the New Science of Bereavement Tells Us About Life After Loss* and *The End of Trauma: How the New Science of Resilience Is Changing How We Think About PTSD*, both by George Bonnano, are good resources for those experiencing these conditions.
- The Centers for Disease Control and Prevention's website (www.cdc.gov) has extensive information on adverse childhood experiences and their effects on health and other important psychosocial outcomes.

References

1. World Health Organization. (2022, June). Mental disorders. www.who.int/newsroom/fact-sheets/detail/mental-disorders.
2. GBD Mental Disorders Collaborators. (2022). Global, regional, and national burden of 12 mental disorders in 204 countries and territories, 1990–2019: A systematic analysis for the Global Burden of Disease Study 2019. *Lancet Psychiatry*, 9, 137–150.
3. Nochaiwong, S., Ruengorn, C., Thavorn, K., Hutton, B., Awiphan, R., Phosuya, C., et al. (2021). Global prevalence of mental health issues among the general population during the coronavirus disease–2019 pandemic: A systematic review and meta-analysis. *Nature Scientific Reports*, 11, Article 10173.
4. Substance Abuse and Mental Health Services Administration. (2022). Key substance use and mental health indicators in the United States: Results from the 2021 National Survey on Drug Use and Health. Center for Behavioral Health Statistics and Quality, Substance Abuse and Mental Health Services Administration. www.samhsa.gov/data/report/2021-nsduh-annual-national-report.
5. American Psychological Association. (2023). Stress in America 2023: A nation recovering from collective trauma. www.apa.org/news/press/releases/stress/2023/collective-trauma-recovery.
6. American Psychological Association. (2022, February 1). Written testimony of Mitch Prinstein, PhD, ABPP, Chief Science Officer, American Psychological Association. Mental health and substance use disorders: Responding to the growing crisis. Before the U.S. Senate Committee on Health, Education, Labor, and Pensions. www.apa.org/news/press/releases/2022/02/testimony-prinstein-mental-substance-disorders.pdf.
7. American Psychiatric Association. (2022). *Diagnostic and statistical manual of mental disorders* (5th ed., Text Revision). American Psychiatric Association Publishing.
8. American Psychiatric Association. (2022). Anxiety disorders. In: *Diagnostic and statistical manual of mental disorders* (5th ed., Text Revision). American Psychiatric Association Publishing.
9. American Psychiatric Association. (2022). Depressive disorders. In: *Diagnostic and statistical manual of mental disorders* (5th ed., Text Revision). American Psychiatric Association Publishing.
10. Mokhtari, S., Mokhtari, A., Bakizadeh, F., Moradi, A., & Shalbafan, M. (2023). Cognitive rehabilitation for improving cognitive functions and reducing the

severity of depressive symptoms in adult patients with major depressive disorder: A systematic review and meta-analysis of randomized control clinical trials. *BMC Psychiatry*, 23, Article 77.

11. Klaus, F., & Eyler, L.T. (2023). The neuropsychology of mood disorders. In: G.G. Brown, T.Z. King, K.Y. Haaland, & B. Crosson (Eds.). *APA handbook of neuropsychology: Vol. 1. Neurobehavioral disorders and conditions: Accepted science and open questions*. American Psychological Association.

12. Aupperle, R.L., McDermott, T.J., White, E., & Kirlic, N. (2023). The neuropsychology of anxiety: An approach-avoidance decision-making framework. In: G.G. Brown, T.Z. King, K.Y. Haaland, & B. Crosson (Eds.). *APA handbook of neuropsychology: Vol. 1. Neurobehavioral disorders and conditions: Accepted science and open questions*. American Psychological Association.

13. Judah, M.R., Grant, D., Mills, A.C., & Lechner, W.V. (2013). The neural correlates of impaired attentional control in social anxiety: An ERP study of inhibition and shifting. *Emotion*, 13(6), 1096–1106.

14. White, E.J., Grant, D.M., Taylor, D.L., Frosio, K.E., Mills, A.C., & Judah, M.R. (2018). Examination of evaluative threat in worry: Insights from the error-related negativity (ERN). *Psychiatry Research: Neuroimaging*, 282, 40–46.

15. American Psychiatric Association. (2022). Trauma- and stressor-related disorders. In: *Diagnostic and statistical manual of mental disorders* (5th ed., Text Revision). American Psychiatric Association Publishing.

16. Ross, D.A., Arbuckle, M.R., Travis, M.J., Dwyer, J.B., van Schalkwyk, G.I., & Ressler, K.J. (2017). An integrated neuroscience perspective on formulation and treatment planning for posttraumatic stress disorder: An educational review. *JAMA Psychiatry*, 74(4), 407–415.

17. Scott, J.C., Lynch, K.G., Cenkner, D.P., Kehle-Forbes, S.M., Polusny, M.A., Gur, R.C., et al. (2021). Neurocognitive predictors of treatment outcomes in psychotherapy for comorbid PTSD and substance use disorders. *Journal of Counseling and Clinical Psychology*, 89(11), 937–946.

18. Punski-Hoogervorst, J.L., Engel-Yeger, B., & Avital, A. (2023). Attention deficits as a key player in the symptomatology of posttraumatic stress disorder: A review. *Journal of Neuroscience Research*, 101, 1068–1085.

19. Grummit, L.R., Kreski, N.T., Kim, S.G., Platt, J., Keyes, K.M., & McLaughlin, K.A. (2021). Association of childhood adversity with morbidity and mortality in US adults: A systematic review. *JAMA Pediatrics*, 175(12), 1269–1278.

20. Barisa, M.T. (2020). Mood disorders: Depression, mania, and anxiety. In: K.J. Stucky, M.W. Kirkwood, & J. Donders (Eds.). *Clinical neuropsychology study guide and board review* (2nd ed.). Oxford University Press.

21. Drysdale, A.T., Grosenick, L., Downar, J., Dunlop, K., Mansouri, F., Meng, Y., et al. (2017). Resting-state connectivity biomarkers define neurophysiological subtypes of depression. *Nature Medicine*, 23(1), 28–43.

22. Ng, T.H., Alloy, L.B., & Smith, D.V. (2019). Meta-analysis of reward processing in major depressive disorder reveals distinct abnormalities within the reward circuit. *Translational Psychiatry*, 9(1), 293.

23. MacNamara, A., & Proudfit, G.H. (2014). Cognitive load and emotional processing in generalized anxiety disorder: Electrocortical evidence for increased distractibility. *Journal of Abnormal Psychology*, 123(3), 557–565.

24. Guerra, L.T.L., Rocha, J.M., Osório, F.L., Bouso, J.C., Hallak, J.E.C., & dos Santos, R.G. (2023). Biases in affective attention tasks in posttraumatic stress

disorder patients: A systematic review of neuroimaging studies. *Biological Psychology*, 183, Article 108660.

25. Bauer, R.M., & Gaynor, L.S. (2020). Functional neuroanatomy and essential neuropharmacology. In: K.J. Stucky, M.W. Kirkwood, & J. Donders (Eds.). *Clinical Neuropsychology Study Guide and Board Review* (2nd ed.). Oxford University Press.
26. Beauregard, M. (2014). Functional neuroimaging studies of the effects of psychotherapy. *Dialogues in Clinical Neuroscience*, 16(1), 75–81.
27. Gelenberg, A.J., Freeman, M.P., Markowitz, J.C., Rosenbaum, J.F., Thase, M.E., Trivedi, M.H., & Van Rhoads, R.S. (2010). *Practice guideline for the treatment of patients with major depressive disorder* (3rd ed.). American Psychiatric Association.
28. Noetel, M., Sanders, T., Gallardo-Gómez, D., Taylor, P., del Pozo Cruz, B., van den Hoek, D., Smith, J.J., Mahoney, J., Spathis, J., Moresi, M., Pagano, R., Pagano, L., Vasconcellos, R., Arnott, H., Varley, B., Parker, P., Biddle, S., & Lonsdale, C. (2020). Effect of exercise for depression: Systematic review and network meta-analysis of randomized controlled trials. *BMJ*, 384, Article e075847.
29. Campbell, K.L., Winters-Stone, K.M., Wiskemann, J., May, A.M., Schwartz, A.L., Courneya, K.S., Zucker, D.S., Matthews, C.E., Ligibel, J.A., Gerber, L.H., Morris, G.S., Patel, A.V., Hue, T.F., Perna, F.M., & Schmitz, K.H. (2019). Exercise guidelines for cancer survivors: Consensus statement from International Multidisciplinary Roundtable. *Medicine & Science in Sports & Exercise*, 51(11), 2375–2390.
30. Grey, I., Arora, T., Thomas, J., Saneh, A., Tohme, P., & Abi-Habib, R. (2020). The role of perceived social support on depression and sleep during the COVID-19 pandemic. *Psychiatry Research*, 293, Article 113452.

6   Medical Conditions

In an article describing her experience with chronic autoimmune disease and the challenges of obtaining proper diagnosis and treatment, author Meghan O'Rourke describes:

> One morning in March, I sat down at my desk to work, and I found I could no longer write or read; my brain seemed enveloped in a thick gray fog. I wondered if it was a result of too much Internet surfing, and a lack of will power. I wondered if I was depressed. But I wanted to work. I didn't feel apathy, only a weird sense that my mind and my body weren't synched. Was I going mad?[1]

A wide range of patients living with certain conditions—from menopause to migraine, those with diabetes and on dialysis, those complaining of "chemobrain" and "fibro fog"—describe significant cognitive changes associated with their medical problems and their treatment. Some of them display impairment on cognitive testing; many perform within normal limits on tests but their lapses in daily life are disruptive and disturbing to them. This chapter reviews how common medical conditions, and the medications used to treat them, can drain our cognitive resources.

* * *

Globally, about one in three adults live with multiple chronic conditions, and the number is predicted to increase significantly in the next decade.[2] In the U.S., about two-thirds of adults have at least one chronic illness.[3] Over one in four live with multiple chronic conditions, but that proportion rises with age and is closer to two-thirds of adults ages 65 or older.[4] High blood pressure (hypertension) and high cholesterol are the most common conditions, and a third of adults have *metabolic disorder*, defined as having three or more of the following: abdominal obesity, high blood pressure, high blood sugar levels, high levels of triglycerides (a type of lipids), and low HDL cholesterol.[5] Metabolic syndrome increases the risk for many other conditions, including coronary heart disease, heart failure, immune

DOI: 10.4324/9781003409311-9

problems, organ damage, sleep apnea, some forms of cancer, stroke, diabetes, and cognitive decline.

In public education campaigns regarding the importance of the prevention, diagnosis, and management of chronic illnesses, the prevalence of cognitive complaints is rarely mentioned. Among adults with two or more chronic medical conditions, 30 to 40 percent report concentration and memory difficulties.[6] Among older adults, as many as 60 percent of those with a chronic illness report cognitive complaints; the more illnesses they live with, the more likely the presence of cognitive complaints.[7] This makes sense given associated issues including disrupted sleep, stress (including healthcare-related financial stress), the need for multiple medications with potential cognitive side effects, and difficulty engaging in physical activity. Even when the cognitive changes are subtle, they can be noticeable to the person and disruptive in their everyday life. Especially if not well controlled, some of these conditions can also increase the risk for late-life cognitive decline.

Cognitive Changes in Medical Conditions

Conditions Involving Organ Failure

Patients with advanced organ disease or organ failure can experience profound cognitive impairment. Thirty to 40 percent of people with congestive heart failure, for example, display cognitive deficits, most commonly attention and memory changes and slow processing speed; another third actually meet criteria for a diagnosis of mild cognitive impairment.[8], [9], [10] Older patients can experience broader and more profound cognitive changes.

Chronic obstructive pulmonary disease (COPD) refers to conditions like emphysema and chronic bronchitis, caused by damage or chronic inflammation in the lungs and/or airways. About 1 in 3 patients with COPD display some degree of cognitive impairment, and 1 in 4 meet criteria for mild cognitive impairment.[8] Decline can be broad, involving attention, memory, executive functions, and processing speed, but cognitive functioning can be preserved in those whose condition is well managed with oxygen therapy.[11]

Advanced liver disease can cause pronounced cognitive and behavioral symptoms.[12] The term *encephalopathy* means broad impairment in brain functioning, usually characterized by confusion and difficulty remaining fully conscious.[13] *Hepatic encephalopathy* (HE), caused by advanced liver disease and its complications, occurs in as many as half of patients with cirrhosis.[14] Symptoms of HE can be mild, with declines in attention, working memory, and processing speed, but HE can progressively worsen and result in behavioral changes (like irritability), confusion, disorientation (not knowing where they are or what day it is), and coma. Episodes of HE are overall reversible with treatment, but patients with severe and poorly managed liver disease who experience multiple episodes can develop chronic cognitive impairment. (Patients with advanced liver disease who were being considered for liver transplants were among the most cognitively impaired I have ever worked with, to the point that some of them were unable to complete cognitive tests at all.) Some cognitive changes have also been documented with milder liver conditions, like non-alcoholic "fatty liver" diseases.[15] While mild, declines tend to be broad, and most notable in attention, cognitive flexibility, and processing speed.

Cognitive complaints are also common among patients with advanced kidney disease.[16], [17], [18], [19] The prevalence of cognitive symptoms increases with the duration and severity of the renal dysfunction and its complications, and can be as high as 70 percent among patients on hemo-dialysis. Patients on hemodialysis can also experience noticeable fluctuations in their cognitive status: Their cognitive problems worsen in the periods in between treatments, and improve after dialysis. In contrast, cognitive functioning seems to be more stable on peritoneal dialysis. Cognitive deficits are usually broad but most prominent on executive functions.

Other Chronic Medical Conditions

Diabetes can cause fluctuations in blood glucose levels, and both hypogly-cemia (abnormally low blood glucose levels) and hyperglycemia (abnor-mally high blood glucose levels) can interfere with cognitive function.[20] As a group, individuals with type 2 diabetes display slower processing speed and poorer attention, memory, and executive performance than adults without diabetes, although the differences are small. Cognitive deficits are more likely in older patients and those with worse glycemic control.

Individuals with hypertension largely perform within normal limits on cognitive tests, but as a group they can perform lower than individuals without hypertension on measures of executive function and processing speed.[21], [22] Like the fluctuations in blood sugar seen in diabetes, fluc-tuations in blood pressure can also cause temporary, reversible cognitive changes, like episodes of mild confusion or mental slowness.

Systemic lupus erythematosus (SLE) is a chronic *autoimmune disease,* a condition in which the body's immune system targets the healthy cells in our organs and tissues. The prevalence of cognitive impairment in SLE can be as high as 95 percent in patients with the neuropsychiatric form of SLE, which affects the central nervous system.[23] Patients with neuropsychiatric SLE can display broad cognitive deficits in processing speed, attention, executive functions, memory, and rapid word retrieval.

Lyme disease is caused by a bacterial infection transmitted by ticks. While most people who receive adequate treatment for the infection recover, 10 to 15 percent develop persistent or recurrent symptoms referred to as *post-treatment Lyme disease syndrome.* As many as 90 percent of them report cognitive complaints, and about half report moderate to severe cognitive changes.[24] However, on formal testing, only a quarter display cognitive decline, and most patients—even those who do experience some decline—perform within the normal range. Cognitive problems seem to be largely due to executive changes and slower processing speed, although attention problems and decreased mental flexibility have also been documented.

Most recently, *post-acute Covid-19 syndrome* (PACS) or "long Covid" has received quite a bit of attention. The World Health Organization has defined long Covid as symptoms that are present three months after infection, with a minimum duration of two months, and that cannot be explained by another diagnosis.[25] A large multinational study found that 6 percent of individuals who had experienced a symptomatic Covid infection were experiencing fatigue, cognitive problems, or respiratory symptoms three months after infection; however, other studies have reported rates of long Covid as high as 40 percent, with higher prevalence among women, those who had to be hospitalized for their initial Covid infection, and those with pre-existing medical conditions.[26] Fatigue is the most common complaint, followed by memory problems. Concentration problems, confusion, "brain fog," anxiety, and depression are also among the most common symptoms reported.

A definition of *post-Covid cognitive dysfunction* has been proposed, as new onset of cognitive impairment present at least three months after Covid infection, with deficits in attention and processing speed; memory and executive deficits may or may not be present.[27] Prevalence estimates vary from 20 to 40 percent of patients who experienced a Covid infection, again with higher prevalence for those who required hospitalization. While structural and functional brain changes have been documented, the cognitive complaints and other symptoms can resolve, and it seems that only about 15 percent of patients continue to experience symptoms a year after their infection. This is a rapidly evolving area of research, with wide variability in findings and much left to understand, including whether long-term symptoms are more common after Covid than other viral infections.

Myalgic encephalomyelitis/chronic fatigue syndrome (ME/CFS) is characterized by unrefreshing sleep, fatigue, and malaise after physical exertion, often accompanied by cognitive complaints and/or *orthostatic intolerance*—the development of autonomic symptoms like palpitations, light-headedness, or dizziness, usually when upright.[28] Cognitive complaints are common and often severe, but on testing group differences between patients and healthy adults tend to be small. Slow processing speed is the most common finding, but lower performance in attention, working memory, and memory tests has also been found.[29] In general, the severity of cognitive symptoms increases with the overall severity of other symptoms.

While not a chronic disease, it is worth mentioning *perimenopause*, the period of time prior to the last menstrual period, characterized by menstrual cycle variability. Perimenopause can last anywhere from a couple of years to a decade, and two-thirds of women report cognitive changes including concentration, memory, and word-finding problems.[30] Some studies have indeed documented mild declines in attention and memory during perimenopause, but performance remains within normal limits. Depression, anxiety, physical symptoms, and sleep disturbance are all associated with the presence and extent of cognitive complaints, and, importantly, cognitive complaints seem to be more pronounced in perimenopausal than postmenopausal women, suggesting that cognitive functioning stabilizes after menopause.

Chronic Pain Conditions

Multiple brain areas are involved in the processing, perception, and regulation of pain.[31] In addition, pain requires us to exert significant inhibition in order to focus on tasks. As with mental health conditions, if we are in pain, we are running the "pain app" in our brain, so in some sense we are constantly multitasking. The prefrontal cortex, specifically, plays an important role in the regulation of pain, how we think about pain sensations, and our emotional response to pain, so pain consumes valuable executive resources.[32] Studies with patients with chronic pain have documented abnormalities in the volume and activation of brain regions involved in pain processing and modulation, and many of these regions—like limbic and prefrontal areas—are also crucial for cognition.[31], [33]

Not surprisingly, patients with chronic pain often report quite prominent cognitive problems. However, studies examining cognitive performance in conditions involving chronic pain—fibromyalgia, migraine, chronic back pain, rheumatoid arthritis, diabetic neuropathy, osteoarthritis, and others— have produced inconclusive results in terms of deficits on cognitive testing: Many do document lower performance on measures of global cognitive functioning, attention, executive functions, and memory, but the group differences between pain patients and healthy adults tend to be small.

Fibromyalgia is a condition characterized by chronic and widespread pain throughout the body, often accompanied by fatigue and sleep disturbance.[34] Cognitive complaints are present in about 75 percent of patients and tend to be diffuse; problems concentrating and thinking clearly are common and often described by the term "fibrofog." While results on testing are inconsistent, decreased performance has been documented across multiple cognitive domains. As in ME/CFS, the extent of cognitive changes seems to be related to the overall severity of other symptoms.

It is well known that migraine can cause temporary cognitive deficits, like concentration problems and word-finding difficulties, before and during the headache phase.[35] As a group, patients with migraine perform slightly lower than adults without migraine across multiple cognitive domains, most significantly in processing speed, complex attention, and executive functions. Like other individuals living with pain conditions, migraine sufferers often show atypical patterns of activity in brain regions involved in pain modulation, but, in addition, white matter lesions and decreased blood flow to certain brain regions have also been documented.

* * *

These medical conditions affect cognition through multiple mechanisms, but two important pathways are cerebrovascular and inflammatory processes. *Cerebrovascular* changes disrupt the supply of oxygenated blood to the brain—for example, because the heart cannot pump sufficient blood (as in heart failure), because of decreased oxygen levels in the blood (as in COPD), or because of damage to the complex network of blood vessels supplying the brain with blood (which can occur in poorly controlled diabetes, hypertension, kidney disease, and other conditions). *Systemic inflammation* refers to the chronic activation of the immune response in the absence of infection or injury, which includes inflammatory processes in the central nervous system (*neuroinflammation*). Cerebrovascular disease and inflammation are associated with structural and functional changes in brain networks and with cognitive changes and decline over time.

In addition, many of these chronic conditions are often accompanied by other problems that can exacerbate cognitive changes, including emotional symptoms like depressed mood, disrupted sleep from discomfort or pain, significant worry, and hypervigilance about bodily symptoms (Is that a migraine coming? Is this normal fatigue or a fibromyalgia/CFS/lupus flare-up?). Remember that the brain evolved to "budget" and administer our body's resources to ensure that our internal state is conducive to survival and health. When there are disruptions in our body systems, the brain is hard at work monitoring them and attempting to restore balance—that is literally its mission.

Medical Treatments and Cognition

Many commonly used medications can have significant cognitive side effects. The presence and extent of the cognitive changes vary depending on the person's health (e.g., their kidney and liver function) and what other medications they are taking. Older adults are particularly vulnerable to cognitive side effects of medications due to changes in our ability to metabolize and clear substances as we age. Especially among older adults, *polypharmacy*—the regular use of at least five medications—is associated with poorer cognitive performance.[3]

A single dose of *opioid* medication, most commonly prescribed for moderate to severe pain, causes temporary declines in attention, working memory, and memory; people who take higher doses or for a longer time show broader declines in cognitive performance.[31], [36] Again, the effects are more pronounced for older adults, who can develop episodes of confusion.[37] In fact, older adults who take prescription opioids show a mildly accelerated rate of cognitive decline over time and have a higher likelihood of being diagnosed with mild cognitive impairment.[36] The effects are small but cumulative: Each opioid prescription increases the speed of cognitive decline by about 7 percent.

Benzodiazepines, often prescribed as sleep aids and for anxiety, decrease alertness and impair memory, specifically interfering with the formation of new memories, meaning the person might not remember what they did or what was said while under the acute effects of the medication.[38], [39] In addition to concerns about becoming dependent on the medication with prolonged use, long-term benzodiazepine users perform lower than non-users on tests of attention, executive function, memory, and processing speed, with inconsistent improvement after abstinence. Among older adults, long-term users are at higher risk for cognitive decline and dementia.[37], [40] Older adults are also more likely to develop cognitive deficits when using a different class of sleep aids, the so-called "Z-drugs."

Many types of medications have cognitive side effects due to their *anticholinergic* properties: Acetylcholine is a widespread neurotransmitter critical for cognitive functioning, and anticholinergic medications block or inhibit its action. Medications with anticholinergic properties include *some* medications prescribed to treat allergies (antihistamines), depression, psychosis, nausea and vomiting (antiemetics), urinary incontinence (antimuscarinics), Parkinson's disease, abdominal pain (GI antispasmodics), and muscle pain (skeletal muscle relaxants).[37] Risk for cognitive impairment increases if the person is taking multiple medications with anticholinergic properties.

Corticosteroids, often prescribed to treat inflammation and suppress the immune system (e.g., for patients with autoimmune disorders or for patients who have received an organ transplant, to prevent organ rejection) are associated with modest declines in executive functions and memory.[41] Many people also experience pronounced mood changes, including depression or irritability. Certain *antiepileptic* drugs, which are also sometimes prescribed as mood stabilizers for bipolar disorder or for migraine control, can also have significant cognitive effects.

Finally, it is well known that chemotherapy can have profound effects on cognitive functioning, a phenomenon colloquially referred to as "chemobrain." Other cancer treatments, such as hormone therapies and targeted therapies, can also induce cognitive changes, which has led to use of the broader term *cancer-related cognitive impairment*. Mild to moderate cognitive deficits have been documented shortly after chemotherapy, most commonly in processing speed, attention, memory, and executive functions.[42] As many as 75 percent of patients report cognitive symptoms three years after chemotherapy, and some report ongoing problems as far out as a decade. However, only one-fourth to one-third show impairment or decline on cognitive testing.[42], [43] Chemotherapy likely affects the brain through multiple mechanisms, including toxicity to neurons, neuroinflammation, and disrupted neurogenesis. Consistent with this, reductions in gray matter volume, changes in white matter, and changes in activation patterns have been documented after chemotherapy. Patients who perform well on cognitive tests sometimes show increased or more widespread brain activation, suggesting their brains are working harder to achieve adequate performance.

What To Do About It

While diseases involving advanced organ failure can cause severe cognitive impairment, most of the medical conditions reviewed in this chapter typically cause relatively mild cognitive changes. Even when we do experience some decline, our level of functioning tends to remain within normal limits. Those mild cognitive changes become highly disruptive when combined with physical and emotional factors like pain, fatigue, stress, and medication side effects, but we can still benefit from the strategies presented in Part III, which decrease cognitive burden. Ultimately, however, achieving adequate medical control of the underlying illness is crucial to improve our functioning, promote restorative changes in our brain's integrity and functioning, and prevent cerebrovascular damage and long-term cognitive decline.

1. Seek help to stay on top of your health management. Managing a chronic condition can be exhausting, stressful, and expensive, but there are things you can do to increase your chances of being successful. Seek reliable information to educate yourself about your conditions and the benefits, risks, and likely outcomes of available treatments. Identify medical providers that are specialists in your conditions and that take the time to educate you and explain the reasoning behind their recommendations. Adhere to your prescribed treatments, and do not make changes without consulting with your providers.

In addition, I encourage you to consider consulting with a behavioral health specialist if you are having difficulty managing your health. Health psychology is a specialty that helps individuals maintain their health, manage their illnesses, and function the best they can given their health status, by considering not just physiological but also psychological and social factors. There are effective interventions, for example, to help us adhere to sometimes complex and difficult treatments (like adjusting to life on dialysis), make challenging behavioral changes (like changing our diets), manage mental health conditions associated with the illness (like depression and anxiety), and even manage chronic pain—in fact, pain psychology is its own subspecialty. Below are some resources to help you with this.

2. Ask for a medication review if you are concerned about cognitive side effects. Do not stop taking any prescribed medications without consulting with your medical providers. If you take multiple medications (perhaps prescribed by different specialists), if you believe you are experiencing cognitive changes that seem associated with a medication change, and especially if you are an older adult, ask your physician for a medication review. The goal of a medication review is to examine all the medications you are taking to detect potentially problematic side effects or interactions, identify opportunities for *deprescribing*—the process of reducing or stopping a medication under medical supervision—and discuss alternatives to your current regimen.[44] Also, make sure your providers are aware of any supplements you are taking, as those can interact with your prescribed medications.

3. *Tend to the garden.* Regular physical activity is probably our most powerful tool to preserve health and prevent disease, including many of the conditions covered in this chapter. Most patients living with chronic illness experience improvements in their physical, emotional, and cognitive functioning from whatever form of exercise they can safely engage in.[45], [46] However, certain forms of exercise are contraindicated in some chronic medical conditions, so it is critical—especially for those with advanced organ failure, orthopedic or pain conditions, physical frailty or balance problems, and those with ME/CFS—to discuss with their medical providers what kind of exercise they can practice, and ideally consult with a physical therapist or rehabilitation specialist experienced with their condition. Similarly, as we will see in Chapter 7, chronically insufficient sleep can have widespread consequences on pretty much every system in our body, and thus can exacerbate any chronic condition or make it much more difficult to manage it.

Finally, social support is an important predictor of a person's ability to successfully manage their health. People with close, supportive relationships are more likely to sustain behavior changes like abstaining from smoking and exercising regularly, maintain self-care practices, and adhere to their medical treatments.[47] Disease-specific support groups can be another important resource, not just by allowing us to receive emotional support from those going through similar experiences, but by learning practical ways to manage the daily challenges of our condition.

Resources

- The National Institutes of Health National Library of Medicine has a website for patients, Medline Plus (www.medlineplus.gov), with extensive sections on *Health Topics* and *Drugs and Supplements*.
- The American Medical Association (www.ama-assn.org) has a series called *What Doctors Wish Patients Knew*, with information about important health topics by experts.
- The Society of Behavioral Medicine (www.sbm.org) has resources for the public under *Healthy Living*.
- The Society for Health Psychology (www.societyforhealthpsychology. org) has information regarding pain psychology in particular.
- To locate a psychologist with expertise in health psychology, you can use the American Psychological Association's Psychologist Locator (http s://locator.apa.org) or the *Find a Psychologist* feature of the website for the National Register of Health Service Psychologists (www.findap

sychologist.org). It is best to keep your keyword/specialty search broad (e.g., search for "pain" instead of "fibromyalgia"). To find a psychologist who is board certified in clinical health psychology, go to the website of the American Board of Professional Psychology (www.abpp.org) and click on *Directory*. You will be able to enter your location, and under "Board Certified In," select *Clinical Health Psychology*.

- The websites for the U.S. Deprescribing Research Network (www.dep rescribingresearch.org) and Deprescribing.Org (www.deprescribing.org) have resources for patients, including advice on how to talk to your physician about your medications and what specific questions to ask. Again, do *not* stop taking any prescribed medications without consulting with your medical providers: These resources are meant to help you think through any concerns you might have and raise them with your prescribers.

References

1. O'Rourke, Meghan. (2013, August 19). What's wrong with me? *The New Yorker.* www.newyorker.com/magazine/2013/08/26/whats-wrong-with-me.
2. Hajat, C., & Stein, E. (2018). The global burden of multiple chronic conditions: A narrative review. *Preventive Medicine Reports*, 12, 284–293.
3. American Psychological Association. (2023). Stress in America 2023: A nation recovering from collective trauma. www.apa.org/news/press/releases/stress/2023/collective-trauma-recovery.
4. Boersma, P., Black, L.I., & Ward, B.W. (2020). Prevalence of multiple chronic conditions among US adults, 2018. *Preventing Chronic Disease: Public Health Research, Practice, and Policy*, 17, Article E106.
5. National Heart, Lung, and Blood Institute. (Updated 2022, May 18). What is metabolic syndrome?www.nhlbi.nih.gov/health/metabolic-syndrome.
6. Jacob, L., Haro, J.M., & Koyanagi, A. (2019). Physical multimorbidity and subjective cognitive complaints among adults in the United Kingdom: A cross-sectional community-based study. *Nature Scientific Reports*, 9, Article 12417.
7. Hill, N.L., Bhargava, S., Brown, M.J., Kim, H., Bhang, I., Mullin, K., Phillips, K., & Mogle, J. (2021). Cognitive complaints in age-related chronic conditions: A systematic review. *PLOS One*, 16(7), Article e0253795.
8. Yohannes, A.M., Chen, W., Moga, A.M., Leroi, I., & Connolly, M.J. (2017). Cognitive impairment in chronic obstructive pulmonary disease and chronic heart failure: A systematic review and meta-analysis of observational studies. *JAMCDA*, 18, 451e1–451e11.
9. Villringer, A., & Laufs, U. (2021). Heart failure, cognition, and brain damage. *European Heart Journal*, 42, 1579–1581.
10. Hammond, C.A., Blades, N.J., Chaudhry, S.I., Dodson, J.A., LongstrethJr., W.T., Heckbert, S.R., Psaty, B.M., Arnold, A.M., Dublin, S., Sitlani, C.M., Gardin, J.M., Thielke, S.M., Nanna, M.G., Gottesman, R.F., Newman, A.B., & Thacker, E.L. (2018). Long-term cognitive decline after newly diagnosed heart failure: Longitudinal analysis in the CHS (Cardiovascular Health Study). *Circulation: Heart Failure*, 11, Article e004476.

11. Bruce, A.S., Aloia, M.S., & Ancoli-Israel, S. (2013). Neuropsychological effects of hypoxia in medical disorders. In: I. Grant & K.M. Adams (Eds.). *Neuropsychological assessment of neuropsychiatric and neuromedical disorders* (3rd ed.). Oxford University Press.

12. Vilstrup, H., Amodio, P., Bajaj, J., Cordoba, J., Ferenci, P., Mullen, K.D., Weissenborn, K., & Wong, P. (2014). Hepatic encephalopathy in chronic liver disease: 2014 Practice Guideline by the American Association for the Study of Liver Diseases and the European Association for the Study of the Liver. *Journal of Hepatology*, 61(3), 642–659.

13. Loring, D.W. (Ed.). (2015). *INS dictionary of neuropsychology and clinical Äneurosciences* (2nd ed.). Oxford University Press.

14. Rahimi, R.S., Brown, K.A., Flamm, S.L., & Brown, R.S. (2021). Overt hepatic encephalopathy: Current pharmacologic treatments and improving clinical outcomes. *The American Journal of Medicine*, 134, 1330–1338.

15. George, E.S., Sood, S., Daly, R.M., & Tan, S.-Y. (2022). Is there an association between non-alcoholic fatty liver disease and cognitive function? A systematic review. *BMC Geriatrics*, 22, Article 47.

16. Pépin, M., Ferreira, A.C., Arici, M., Bachman, M., Barbieri, M., Bumblyte, I.A., Carriazo, S., Delgado, P., Garneata, L., Giannakou, K., Godefroy, O., Gridzicki, T., Klimkowicz-Mrowiec, A., Kurganaite, J., Liabeuf, S., Mocanu, C.A., Paolisso, G., Spasovski, G., Vazelov, E.S., … Więcek, A., CONNECT Action (Cognitive Decline in Nephro-Neurology European Cooperative Target). (2021). Cognitive disorders in patients with chronic kidney disease: Specificities of clinical assessment. *Nephrology Dialysis Transplantation*, 37, ii23–ii32.

17. Steinbach, E.J., & Harshman, L.A. (2022). Impact of chronic kidney disease on brain structure and function. *Frontiers in Neurology*, 13, Article 979503.

18. Hannan, M., Steffen, A., Quinn, L., Collings, E.G., Phillips, S.A., & Bronas, U.G. (2019). The assessment of cognitive function in older adult patients with chronic kidney disease: An integrative review. *Journal of Nephrology*, 32(2), 211–230.

19. Kelly, D.M., Ademi, A., Doehner, W., Lop, G.Y.H., Mark, P., Toyoda, K., Wong, C.X., Sarnak, M., Cheung, M., Herzog, C.A., Johansen, K.L., Reinecke, H., & Sood, M.M. (2021). Chronic kidney disease and cerebrovascular disease: Consensus and guidance from a KDIGO controversies conference. *Stroke*, 52, e328–e346.

20. Brands, A.M.A., & Kessels, R.P.C. (2013). Diabetes and the Brain: Cognitive performance in type 1 and type 2 diabetes. In: I. Grant & K.M. Adams, (Eds.). *Neuropsychological assessment of neuropsychiatric and neuromedical disorders* (3rd ed.). Oxford University Press.

21. Ungvari, Z., Toth, P., Tarantini, S., Prodan, C.I., Sorond, F., Merkely, B., & Csiszar, A. (2021). Hypertension-induced cognitive impairment: From pathophysiology to public health. *Nature Reviews Nephrology*, 17, 639–654.

22. Iadecola, C., Yaffe, K., Biller, J., Bratzke, L.S., Faraci, F.M., Gorelick, P.B., Gulati, M., Kamel, H., Knopman, D.S., Launer, L.J., Saczynski, J.S., Seshadri, S., & Al Hazzouri, A.Z., on behalf of the American Heart Association Council on Hypertension; Council on Clinical Cardiology; Council on Cardiovascular Disease in the Young; Council on Cardiovascular and Stroke Nursing; Council on Quality of Care and outcomes Research; and Stroke Council (2016). Impact of hypertension on cognitive function: A scientific statement from the American Heart Assocation. *Hypertension*, 68(6), e67–e94.

23. Zabala, A., Salgueiro, M., Sáez-Atxukarro, O., Ballesteros, J., Ruiz-Irastorza, G., & Segarra, R. (2018). Cognitive impairment in patients with neuropsychiatric and non-

neuropsychiatric systemic lupus erythematosus: A systematic review and meta-analysis. *Journal of the International Neuropsychological Society*, 24, 629–639.

24. Touradji, P., Aucott, J.N., Yang, T., Rebman, A.W., & Bechtold, K.T. (2019). Cognitive decline in post-treatment Lyme disease syndrome. *Archives of Clinical Neuropsychology*, 34, 455–465.

25. Global Burden of Disease Long COVID Collaborators. (2023). Estimated global proportions of individuals with persistent fatigue, cognitive, and respiratory symptom clusters following symptomatic COVID-19 in 2020 and 2021. *JAMA*, 328(16), 1604–1615.

26. Chen, C., Haupert, S.R., Zimmerman, L., Shi, X., Fritsche, L.G., & Mukherjee, B. (2022). Global prevalence of post-coronavirus disease 2019 (COVID-19) condition or long COVID: A meta-analysis and systematic review. *The Journal of Infectious Diseases*, 226, 1593–1607.

27. Quan, M., Wang, X., Gong, M., Wang, Q., Li, Y., & Jia, J. (2023). Post-COVID cognitive dysfunction: Current status and research recommendations for high risk population. *The Lancet Regional Health—Western Pacific*, 38, Article 100836.

28. Bateman, L., Bested, A.C., Bonilla, H.F., Chheda, B., Chu, L., Curtin, J.M., Depsey, T.T., Dimmock, M.E., Dowell, T.G., Felsenstein, D., Kaufman, D.L., Klimas, N.G., Komaroff, A.L., Lapp, C.W., Levine, S.M., Montoya, J.G., Natelson, B.H., Peterson, D.L., Podell, R.N., ... Yellman, B.P. (2021). Myalgic encephalomyelitis/chronic fatigue syndrome: Essentials of diagnosis and management. *Mayo Clinic Proceedings*, 96(11), 2861–2878.

29. Joustra, M.L., Hartman, C.A., Bakker, S.J.L., & Rosmalen, J.G.M. (2022). Cognitive task performance and subjective cognitive symptoms in individuals with chronic fatigue syndrome or fibromyalgia: A cross-sectional analysis of the Lifelines Cohort Study. *Psychosomatic Medicine*, 84, 1087–1095.

30. Greendale, G.A., Karlamangla, A.S., & Maki, P.M. (2020). The menopause transition and cognition. *JAMA*, 323(15), 1495–1496.

31. Khera, T., & Rangasamy, V. (2021). Cognition and pain: A review. *Frontiers in Psychology*, 12, Article 673962.

32. Corti, E.J., Gasson, N., & Loftus, A.M. (2021). Cognitive profile and mild cognitive impairment in people with chronic lower back pain. *Brain and Cognition*, 151, Article 105737.

33. Dehghan, M., Schmidt-Wilcke, T., Pfleiderer, B., Eickhoff, S.B., Petzke, F., Harris, R.E., Montoya, P., & Burgmer, M. (2016). Coordinate-based (ALE) meta-analysis of brain activation in patients with fibromyalgia. *Human Brain Mapping*, 37, 1749–1758.

34. Bell, T., Trost, Z., Buelow, M.T., Clay, O., Younger, J., Moore, D., & Crowe, M. (2018). Meta-analysis of cognitive performance in fibromyalgia. *Journal of Clinical and Experimental Neuropsychology*, 40(7), 698–714.

35. Pizer, J.H., Aita, S.L., Myers, M.A., Hawley, N.A., Ikonomou, V.S., Brasil, K.M., Hernandez, K.A., Pettway, E.C., Owen, T., Borgogna, N.C., Smitherman, T.A., & Hill, B.D. (2024). Neuropsychological function in migraine headaches: An expanded comprehensive multidomain meta-analysis. *Neurology*, 102, Article e208109.

36. Warner, N.S., Hanson, A.C., Schulte, P.J., Habermann, E.B., Warner, D.O., & Mielke, M.M. (2022). Prescription opioids and longitudinal changes in cognitive function in older adults: A population-based observational study. *Journal of the American Geriatrics Society*, 70, 3526–3537.

37. American Geriatrics Society. (2023). American Geriatrics Society 2023 updated AGS Beers Criteria® for potentially inappropriate medication use in older adults. *Journal of the American Geriatrics Society, 71,* 2052–2081.
38. Crowe, S.F., & Stranks, E.K. (2018). The residual medium and long-term cognitive effects of benzodiazepine use: An updated meta-analysis. *Archives of Clinical Neuropsychology, 33,* 901–911.
39. Zhang, Y., Zhou, X., Meranus, D.H., Wang, L., & Kukull, W.A. (2016). Benzodiazepine use and cognitive decline in elderly with normal cognition. *Alzheimer's Disease and Associated Disorders, 30*(2), 113–117.
40. Penninkilampi, R., & Eslick, G.D. (2018). A systematic review and meta-analysis of the risk of dementia associated with benzodiazepine use, after controlling for protopathic bias. *CNS Drugs, 32,* 485–497.
41. Prado, C.E., & Crowe, S.F. (2019). Corticosteroids and cognition: A meta-analysis. *Neuropsychology Review, 29,* 288–312.
42. Lange, M., Joly, F., Vardy, J., Ahles, T., Dubois, M., Tron, L., Winocur, G., De Ruiter, M.B., & Castel, H. (2019). Cancer-related cognitive impairment: An update on state of the art, detection, and management strategies in cancer survivors. *Annals of Oncology, 30,* 1925–1940.
43. Whittaker, A.L., George, R.P., & O'Malley, L. (2022). Prevalence of cognitive impairment following chemotherapy treatment for breast cancer: a systematic review and meta-analysis. *Nature Scientific Reports, 12,* Article 2135.
44. O'Mahony, D., Cherubini, A., Guiteras, A.R., Denkinger, M., Beuscart, J-B., Onder, G., Gudmundsson, A., Cruz-Jentoft, A.J., Knol, W., Bahat, G., van der Velde, N., Petrovic, M., & Curtin, D. (2023). STOPP/START criteria for potentially inappropriate prescribing in older people: Version 3. *European Geriatric Medicine, 14,* 625–632.
45. Campbell, K.L., Winters-Stone, K.M., Wiskemann, J., May, A.M., Schwartz, A.L., Courneya, K.S., Zucker, D.S., Matthews, C.E., Ligibel, J.A., Gerber, L.H., Morris, G.S., Patel, A.V., Hue, T.F., Perna, F.M., & Schmitz, K.H. (2019). Exercise guidelines for cancer survivors: Consensus statement from International Multidisciplinary Roundtable. *Medicine & Science in Sports & Exercise, 51*(11), 2375–2390.
46. Ren, X., Wang, X., Sun, J., Hui, Z., Lei, S., Wang, C., & Wang, M. (2022). Effects of physical exercise on cognitive function of breast cancer survivors receiving chemotherapy: A systematic review of randomized controlled trials. *The Breast, 63,* 113–122.
47. Williams, S.L., Haskard-Zolnierek, K.B., & DiMatteo, M.R. (2014). Psychosocial predictors of behavior change. In K.A. Riekert, J.K. Ockene, & L. Pbert (Eds.), *The handbook of health behavior change* (4th ed.). Springer.

7 Sleep

In many professions, insufficient and/or irregular sleep is seen as a necessary sacrifice. Among people who work in production, healthcare, protective service (like law enforcement and firefighting), and food preparation and service, as many as 40 percent sleep less than the recommended amount.[1]

Until a couple of decades ago, medical residents often worked 90 hours of more per week, sometimes in 36-hour shifts with periods of 12 hours or less of rest in between.[2] Proposals to reduce their work hours were met with concerns about compromising the training required to ensure their competence as physicians. In 1984, the issue came under public scrutiny when Libby Zion, an 18-year-old college freshman, died after seeking medical attention in a New York City ER. Her family sued, questioning, among other things, the ability of medical residents working 36-hour shifts to provide appropriate care. While fatigue was not determined to be a direct contributor to her death, the case resulted in the state limiting resident work hours to an average of 80 hours per week, and in 2003 the Accreditation Council for Graduate Medical Education issued nationwide requirements similarly limiting resident workweeks to an 80-hour average.

You do not have to be a physician, surgeon, police officer, or firefighter to face potentially deadly consequences of sleep deprivation. Sometime around 2005, while in graduate school, working on the side, and single-parenting a toddler, I found myself experiencing bouts of severe insomnia. One day, I was stopped at a red light, first in line in the left turn lane. When the light turned green, obviously only for cars going straight in both directions, I automatically started making my left turn—directly into three lanes of incoming traffic. I was able to slam on my breaks in time, but the memory still makes me shudder. This story is not unusual, and yet sleep is often the first thing to go out the window when we are pressed for time, and sometimes we neglect it for no good reason at all. Moreover, all the conditions we have already discussed—stress, depression, and physical problems like pain—disrupt our sleep. Our brains pay the price.

* * *

DOI: 10.4324/9781003409311-10

About 1 in 3 adults in the U.S. sleep, on average, less than the recommended minimum of seven hours of sleep per night, although rates vary significantly by geographic area, from 25 to 50 percent.[3] A third of adults describe their sleep as *very good* or *excellent*, a third as *good*, and a third as *fair* or *poor*.[4] Consistent with the well-known relationship between sleep and mental health, people who rate their mental health highly are six times more likely to have high-quality sleep. More than half of people, particularly women, say getting a good night's sleep is a major priority on weekdays, rated above spending time with family and friends. Stress, depression, anxiety, sleep apnea, and chronic pain all decrease the likelihood of good sleep, while regular exercise and a diet high in produce are associated with better sleep.

Even in the absence of a chronic sleep disorder, insufficient or poor-quality sleep can result in significant cognitive lapses that can be disruptive at best and life-threatening at worst.

Sleep Is Non-Negotiable

Two points are likely sufficient to highlight the importance of sleep. First, chronically poor sleep—either insufficient sleep or sleep of poor quality—is associated with an increased risk for diabetes, hypertension, hyperlipidemia, cardiac and coronary disease, stroke, impaired immune function, chronic pain, cancer, infertility, decreased cognitive performance, increased errors, depression, anxiety, bipolar disorder, Alzheimer's disease, suicidal thoughts, suicide attempts, completed suicide, accidents, and overall mortality risk. [5], [6] Even acutely, a night of poor sleep results in reduced cardiovascular, metabolic, and respiratory functioning. Any physiological process whose disruption causes this breadth of health effects clearly serves vital restorative functions, even if the detailed physiological mechanisms are not fully understood.

Second, we simply cannot override our body's need for sleep: Sleep will eventually win. The body also prioritizes *how* it recovers from sleep loss: After a period of sleep deprivation, we fall asleep faster and more deeply than usual.

Let's back up a bit. You might remember that there are two types of sleep stages, rapid eye movement sleep, or REM sleep, and non-REM sleep. During REM sleep, the stage of sleep during which we have our most vivid dreams, the brain is very active, and physiological measures of brain activity look almost like they do when we are awake. In contrast, the body is largely paralyzed, with the exception of the eye movements that give this stage its name. Non-REM sleep has different stages, from lighter sleep to the deepest, most restorative sleep. This organization of sleep into different stages is called *sleep architecture*.

Every night, under normal circumstances, we cycle through all stages of sleep in predictable patterns. The proportion of time we spend in the different stages changes through the night: We spend more time in deep sleep in the first half of the night, and we spend more time in REM sleep in the second half of the night. Many factors, including sleep deprivation and substance use, alter this normal sleep architecture. Notice that, because the sleep stages are not distributed evenly through the night, sleep disruption can affect one type of sleep more than another, depending, for example, on whether you go to bed too late or wake up too early.[5] The first night after a night of sleep deprivation, we spend more time in deep sleep than in REM sleep; in subsequent nights, we spend more time in REM sleep than usual. While we never quite recover all the sleep we lost, the brain will prioritize the sleep the body needs. Sleep is simply non-negotiable.[7]

As we age, we do not need any less sleep, but we have increasing difficulty obtaining the amount and quality of sleep we need.[8], [9] We sleep less, we spend less time in deep sleep and REM sleep, and our sleep is less efficient and more fragmented—it takes us longer to fall asleep and we wake up more during the night. Our *circadian rhythms* (see Chapter 20) change, and we tend to fall asleep earlier. Sleep disorders, like insomnia and obstructive sleep apnea (OSA), become more common in older age.[10] This is concerning because chronically poor sleep in older age is also associated with cognitive decline: Older adults with ongoing poor sleep quality, abnormalities in circadian rhythms, and sleep disorders like insomnia and OSA have more than one and a half times higher risk of being diagnosed with Alzheimer's disease.[11] Multiple striking studies have documented that brain levels of the proteins characteristic of Alzheimer's disease increase not only with chronic sleep deprivation, but after a single night of sleep loss.[12], [13], [14]

Sleep and Cognition

Sleep seems to allow the brain to metabolize nutrients, prune excess connections between neurons, and clear waste and toxins.[7], [17] Sleep deprivation is associated with reduced metabolic activity in a broad network of regions important for attention, perceptual processing, and executive control, and with increased activation in the default mode network, the network that is active when the brain is idle.

As a general principle, cognitive performance declines the longer we have been awake. When we have been awake more than 16 hours, our alertness fluctuates, and our cognitive performance becomes unreliable.[18] Even relatively mild sleep insufficiency, like sleeping six hours for a couple of nights, results in slowed processing and reaction speeds, more frequent attentional lapses, and increased variability in our cognitive performance. The lapses we experience when sleep deprived are different from other attentional lapses, because they can look like *microsleeps*, during which we actually stop processing sensory input for a few seconds and our eyes might partially or fully close.[18] This is why even mild sleep deprivation can be dangerous: Closing your eyes even for a couple of seconds while driving can have devastating consequences. We are also more likely to make both *omission* and *commission* errors: We miss things that are happening around us and we are more likely to respond when there is nothing to respond to. Again, if you think about driving, it means that you might both miss the small child running towards the street, and slam on your breaks unnecessarily because you think you see something move in the periphery.

Sleep is critical for memory functioning. During non-REM sleep, recent memories that are being held in temporary storage in the hippocampus are consolidated, meaning they are "moved" into longer, more stable storage in the cortex.[5], [15] During non-REM sleep, the brain also weeds out and removes unnecessary neural connections. We basically clean up and refresh our temporary memory storages by archiving what we need to keep and removing what is unnecessary, leaving memory systems ready to learn more information the next day. During REM sleep, the brain reactivates experiences, emotions, and information, integrating them, creating associations between them, and strengthening connections between them. This is important not only for learning, but for creativity and novel problem-solving. You might have had the experience of being "stuck" in a problem, then seeing a novel solution clearly after a good night of sleep.

Anything that interferes with sleep—mood disorders, recreational substances, pain—will interfere with learning and memory. To summarize a large body of literature in very simple terms: If we do not sleep enough before learning, we learn less information the next day, and if we do not sleep after learning, we forget more of what we learned.[19] Importantly, even if we eventually get some recovery sleep, what we forgot from any sleep deprivation we experienced immediately after learning is not recovered—it is lost. In addition, like many other conditions we have reviewed, sleep loss makes our brains work harder. When trying to recall information while sleep deprived, memory networks in temporal regions are less active than normal, while executive networks in prefrontal and parietal areas are overactive, suggesting the need for increased attentional and executive effort and compensation.[18]

The prefrontal cortex seems to be particularly sensitive to the effects of sleep loss, so it makes sense that insufficient sleep affects attention and executive functions. Decreases can be seen in working memory, cognitive inhibition, and cognitive flexibility, although at least some of these deficits are due to sleep processing speed and attention lapses.[18], [20] Consistent with what we reviewed above about REM sleep, creative thinking and novel problem-solving tend to be particularly affected with sleep deprivation. However, aspects of problem-solving that rely more on long-term knowledge and less on executive functions might be more resistant to sleep loss. Let's say you are a very experienced medical provider. Sleep loss might not impact as much your ability to figure out what medications and in what doses a patient might need based on their lab values (an activity that relies on your expert medical knowledge), but you might make mistakes looking up those lab values in a new, user-unfriendly electronic health record system, especially under time pressure and with lots of distractions around (an activity that relies on processing speed, attention, and executive functions).

Some of the most pronounced effects of sleep loss are on our behaviors and emotions. When we are sleep deprived, we become more emotionally reactive to both positive and negative experiences; overall, our mood worsens, it is difficult to have positive thoughts, we are more likely to interpret neutral stimuli negatively, we are less likely to take action to solve problems, we rely more on unproductive coping strategies (like superstitious thinking), we feel more depressed, anxious, and even paranoid, we feel less empathy, we are worse at understanding interpersonal dynamics, we have more physical complaints, we are less able to delay gratification, we are more impulsive, we take more risks, we are more sedentary, we are more likely to use substances, we crave more carbohydrates, and we are more likely to relapse if we have a substance use problem.[5], [7], [18] These effects make sense given that sleep deprivation affects the functioning of executive networks that help us plan and direct our behavior based on goals (like being healthy), suppresses some of the connectivity between prefrontal areas and limbic areas (thus decreasing our ability to regulate our emotions), and results in neurochemical changes in systems including those that guide our behavior towards pleasure and rewards.

To put these effects into context: After being awake for 19 hours, our cognitive performance is equivalent to that of a person with 0.08 percent blood alcohol level, the legal limit for driving under the influence.[5] Moreover, these effects can be cumulative: If we sleep six hours a night for ten days, our cognitive performance looks like that of a person who has gone without sleep for 24 hours straight.

Cognition in Sleep Disorders

The most common sleep disorder, insomnia, refers to insufficient quantity or quality of sleep due to ongoing difficulty falling asleep, difficulty staying asleep through the night, or waking up too early in the morning.[21] Notice that in insomnia, the person is unable to sleep long enough or well enough despite adequate opportunity to sleep. This diagnosis does not apply to those of us who get insufficient sleep because we stay up late watching cat videos or working. The person with insomnia cannot sleep enough or well enough despite trying to fall asleep, often trying for a long period of time. Insomnia is associated with poorer overall cognitive performance, with most pronounced declines in complex attention, episodic memory, and executive functions including working memory and problem-solving.[22], [23]

Obstructive sleep apnea is a disorder caused by the repetitive collapse of the upper airway during sleep, causing periods when breathing stops (apnea) or is reduced (hypopnea).[21], [24] It is associated with sleep fragmentation (e.g., from awakenings due to gasping), loud snoring, low blood oxygen levels (hypoxemia), and high CO_2 blood levels (hypercapnia). OSA can cause declines in attention, executive functioning, processing speed, and visuospatial abilities.[9] It is a significant risk factor for cognitive decline, presenting a higher risk than other sleep disorders.[11] Brain changes have been documented, especially when OSA is severe and not controlled, including loss of volume in areas that are part of memory and executive networks, loss of white matter integrity, and neuronal damage and loss in the hippocampus.[9], [25] OSA significantly increases the risk of stroke, white matter ischemic damage, and small bleeds and *infarctions* (areas where brain tissue has died due to inadequate blood supply). Fortunately, treatment with continuous positive airway pressure (CPAP) can result in improved cognitive functioning, restorative changes in white matter (possibly through reduced inflammation and improved blood flow), slower rates of cognitive decline, and reduced risk for mild cognitive impairment and dementia.[25], [26]

What To Do About It

Let me say it again: Sleep is non-negotiable. And if you are already depleted and overwhelmed, dealing with chronic stress, illnesses, or mood problems, it is even more crucial to make it your priority.

1. Pay attention to your sleep and sleepiness. Adults should sleep seven or more hours per night on a regular basis.[6] If you have been sleep deprived for some time, you might not realize anymore just how sleep deprived you are. Symptoms that suggest the presence of sleep deprivation or a sleep disorder include:

- Excessive daytime sleepiness. How likely are you to nod off if you are sitting down watching TV or reading? If you're riding in a car for an hour? If you're stopped in traffic for a few minutes? (Notice that sleepiness is not fatigue: You can be tired and feel like you need to sit down or lie down to rest, but if you are not sleepy, you will not nod off.)
- Snoring, choking, or gasping that wakes you up.
- Waking up feeling unrested.
- Waking up with a headache.

You can use a simple version of a sleep diary (see *Resources*), an app, or a monitoring device, to gather some data. Keep in mind that the accuracy of many sleep-monitoring devices and apps, however, has not been established. If you decide to seek medical evaluation for a possible sleep disorder, you can use the resources listed below to identify a specialist. Luckily, we have effective treatments for sleep disorders. The first-line treatment for insomnia is Cognitive Behavioral Therapy for Insomnia (CBT-I), proven to be more effective than medications.[27] There are other effective behavioral therapies, including brief treatments. For OSA, positive airway pressure is the treatment of choice, but other effective therapies (oral appliances, surgical procedures, and behavioral strategies) are available.[24]

There are many considerations regarding the use of pharmacological sleep aids, from prescription medications to supplements, including side effects, potential for abuse and dependence, safety concerns, interactions, short-term or minimal efficacy, etc. Do not start using a pharmacological sleep aid without consulting at least with your primary care physician, and ideally with a sleep medicine specialist.

2. Protect your sleep fiercely. In the absence of a sleep disorder, there are many things you can do to optimize your sleep.[5]

- Get enough sleep. Remember that to ensure you sleep at least seven hours, you will need to spend at least seven and a half hours in bed. Do not wait until 11 p.m. to head to bed if you have to wake up at 6 a.m.
- Go to bed and wake up at a fixed hour, even on weekends.
- Develop a wind-down routine: Spend the last half-hour or hour before going to bed signaling to your brain it is time to sleep. Turn your devices off. Play relaxing music. Read. Write in a journal to get anxious thoughts out of your head.
- Avoid exercising in the 2–3 hours before bedtime.
- Schedule your meals so you do not go to bed right after eating, or when you are hungry.
- Artificial light—from lighting but also laptop and phone screens—can inhibit the release of melatonin that should naturally happen in the evenings, making it more difficult to fall asleep. It also reduces REM sleep. Limit your exposure to light in the hours before bed, avoid electronic devices for at least an hour before bedtime, and keep your bedroom as dark as safely possible.
- Keep your night-time temperature low. Our core temperature decreases before we fall asleep. Keep your bedroom temperature just a bit lower than you might prefer. Take a warm bath (this might seem counterintuitive, but it helps because when you get out, your body releases the heat and your core body temperature drops).
- Make your bed a cue for sleep. Avoid doing things in bed that are incongruent with sleepiness, like watching videos on your phone, and avoid sleeping elsewhere, like the couch.
- Do not lie awake in bed for too long. If you cannot fall asleep or if you wake up in the middle of the night and cannot go back to sleep, get out of bed and do something relaxing until you feel you might be able to fall asleep. Do not stare at the clock!
- Avoid napping after 3 p.m.
- Avoid drinking caffeine in the afternoon. Half of the caffeine you consume will still be in your system 5–7 hours after you drink it.

- Avoid drinking alcohol before bed. It might feel like it helps you fall asleep easily, but it causes night-time awakenings (that you might not even remember) and it suppresses REM sleep.
- Avoid having too much to drink close to bedtime, so you do not wake up to use the restroom.
- Spend time in natural sunlight in the mornings. This can help regulate your sleep patterns. If you are an older adult who is falling asleep too early, you can try getting some sunlight exposure in the afternoon.

3. Tend to the garden. Every condition we are reviewing in Part II—stress, psychological and medical conditions, substance use, aging—can negatively affect sleep. And poor sleep can exacerbate almost every condition—it can worsen our stress, our psychological symptoms, our pain, and chronic diseases. It makes it less likely that we will exercise, eat healthily, or adhere to our medical regimen. The relationship between our sleep health and other physical and mental health problems can become a vicious cycle—for example, we can develop insomnia after a stressful life event, such as losing our job, but insomnia itself can then lead to further stressful events, perhaps through impaired decision-making, daytime sleepiness, fatigue, and emotion dysregulation.[28] Think about it: If you are going through life sleep deprived, you might miss due dates and deadlines, make impulsive decisions, and find yourself in more interpersonal conflicts due to irritability and poor emotion regulation, all of which can result in even more stressors.

The good news is that the relationship between sleep and every other aspect of our health also works to our advantage: By improving our health behaviors, we can improve our sleep. When we exercise routinely, for example, we fall asleep faster, sleep more, spend more time in deep sleep, have better sleep quality, and wake up less during the night. And by improving our sleep, we can improve basically every aspect of our health. Sleep truly is the closest thing we have to a royal road to a brain-friendly life.

Resources

- The book *Why We Sleep: Unlocking the Power of Sleep and Dreams*, by Matthew Walker, Ph.D., is an excellent, enjoyable, and accessible book regarding sleep and its critical role in health and disease.
- The American Academy of Sleep Medicine has a public education website (www.sleepeducation.org) with extensive education on sleep health, sleep disorders, and their management. It also has links to tools like self-assessments and sleep diaries.
- Two other websites with helpful information, including information about CBT-I, are the National Sleep Foundation (www.thensf.org) and the Sleep Foundation (www.sleepfoundation.org).

- To find a behavioral sleep medicine specialist, you can search the directories of the Society of Behavioral Sleep Medicine (www.behavioralsleep.org) and the American Board of Sleep Medicine (www.absm.org, under "Credential Verification").

References

1. Centers for Disease Control and Prevention. (2017). Short sleep duration by occupation group—29 states, 2013–2014. U.S. Department of Health and Human Services. www.cdc.gov/mmwr/volumes/66/wr/mm6608a2.htm.
2. Ulmer, C., Wolman, D.M., & Johns, M.M.E. (Eds). (2009). *Resident duty hours. Enhancing sleep, supervision, and safety.* Institute of Medicine of the National Academies. The National Academies Press.
3. National Center for Chronic Disease Prevention and Health Promotion, Division of Population Health. (2022, November 2). FastStats: Sleep in Adults. Centers for Disease Control and Prevention. www.cdc.gov/sleep/data-research/facts-stats/adults-sleep-facts-and-stats.html.
4. Gallup. (2022). *The state of sleep in America: 2022 report.* Gallup.
5. Walker, M. (2017). *Why we sleep. Unlocking the power of sleep and dreams.* Scribner.
6. Watson, N.F., Badr, M.S., Belenky, G., Bliwise, D.L., Buxton, O.M., Buysse, D., Dinges, D.F., Gangwisch, J., Grandner, M.A., Kushida, C., Malhotra, R.K., Martin, J.L., Patel, S.R., Quan, S.F., & Tasali, E. (2015). Recommended amount of sleep for a healthy adult: A joint consensus statement of the American Academy of Sleep Medicine and Sleep Research Society. *Sleep,* 38(6), 843–844.
7. Grandner, M.A., & Fernandez, F.X. (2021). The translational neuroscience of sleep: A conceptual framework. *Science,* 374(6567), 568–573.
8. Hokett, E., Arunmozhi, A., Campbell, J., Verhaeghen, P., & Duarte, A. (2021). A systematic review and meta-analysis of individual differences in naturalistic sleep quality and episodic memory performance in young and older adults. *Neuroscience and Biobehavioral Reviews,* 127, 675–688.
9. Wennberg, A.M.V., Wu, M.N., Rosenberg, P.B., & Spira, A.P. (2017). Sleep disturbance, cognitive decline, and dementia: A review. *Seminars in Neurology,* 37(4), 395–406.
10. Dzierzewski, J., Dautovich, N., & Ravyts, S. (2018). Sleep and cognition in older adults. *Sleep Medicine Clinics,* 13, 93–106.
11. Bubu, O.M., Brannick, M., Mortimer, J., Umasabor-Bubu, O., Sebastião, Y.V., Wen, Y., Schwartz, S., Borenstein, A.R., Wu, Y., Morgan, D., & Anderson, W.M. (2017). Sleep, cognitive impairment, and Alzheimer's disease: A systematic review and meta-analysis. *Sleep,* 40(1), 1–18.
12. Holth, J.K., Fritschi, S.K., Wang, C., Pedersen, N.P., Cirrito, J.R., Mahan, T.E., Finn, M.B., Manis, M., Geerling, J.C., Fuller, P.M., Lucey, B.P., & Holtzman, D.M. (2019). The sleep-wake cycle regulates brain interstitial fluid tau in mice and CSF tau in humans. *Science,* 363, 880–884.
13. Ooms, S., Overeem, S., Besse, K., Rikkert, M.O., Verbeek, M., & Claassen, J.A.H.R. (2014). Effect of 1 night of total sleep deprivation on cerebrospinal fluid ß-amyloid 42 in healthy middle aged men: A randomized clinical trial. *JAMA Neurology,* 71(8), 971–977.

14. Shokri-Kojori, E., Wang, G-J., Wiers, C.E., Demiral, S.B., Guo, M., Kim, S.W., Lindgren, E., Ramirez, V., Zehra, A., Freeman, C., Miller, G., Manza, P., Srivastava, T., De Santi, S., Tomasi, D., Benveniste, H., & Volkow, N.D. (2018). ß-amyloid accumulation in the human brain after one night of sleep deprivation. *PNAS*, 115(17), 4483–4488.
15. Mantua, J., & Spencer, R.M.C. (2017). Exploring the nap paradox: Are mid-day sleep bouts a friend or foe? *Sleep Medicine*, 37, 88–97.
16. Ma, Y., Liang, L., Zheng, F., Shi, L., Zhong, B., & Xie, W. (2020). Association between sleep duration and cognitive decline. *JAMA Network Open*, 3(9), Article e2013573.
17. Siegel, J.M. (2022). Sleep function: An evolutionary perspective. *Lancet Neurology*, 21(10), 937–946.
18. Killgore, W.D.S. (2010). Effects of sleep deprivation on cognition. In: G.A. Kerkhof & H.P.A. Van Dongen (Eds.). *Progress in Brain Research*, Vol. 185. Elsevier.
19. Leong, R.L.F., & Chee, M.W.L. (2023). Understanding the need for sleep to improve cognition. *Annual Review of Psychology*, 74, 27–57.
20. Brownlow, J.A., Miller, K.E., & Gehrman, P.R. (2020). Insomnia and cognitive performance. *Sleep Medicine Clinics*, 15, 71–76.
21. American Psychiatric Association. (2022). Sleep-wake disorders. In *Diagnostic and statistical manual of mental disorders* (5th ed., Text Revision). American Psychiatric Association Publishing.
22. Wardle-Pinkston, S., Slavish, D.C., & Taylor, D.J. (2019). Insomnia and cognitive performance: A systematic review and meta-analysis. *Sleep Medicine Reviews*, 48, Article 101205.
23. Fortier-Brochu, E., Beaulieu-Bonneau, S., Ivers, H., & Morin, C.M. (2012). Insomnia and daytime cognitive performance: A meta-analysis. *Sleep Medicine Reviews*, 16, 83–94.
24. Epstein, L.J., Kristo, D., Strollo, P.J., Friedman, N., Malhotra, A., Patil, S.P., Ramar, K., Rogers, R., Schwab, R., Weaver, E.M., & Weinstein, M.D. (2009). Clinical guideline for the evaluation, management, and long-term care of obstructive sleep apnea in adults. *Journal of Clinical Sleep Medicine*, 5(3), 263–276.
25. Shieu, M.M., Zaheed, A.N., Shannon, C., Chervin, R.D., Conceicao, A., Paulson, H.L., Braley, T.J., & Dunietz, G.L. (2022). Positive airway pressure and cognitive disorders in adults with obstructive sleep apnea: A systematic review of the literature. *Neurology*, 99, e334–e346.
26. Pollicina, I., Maniaci, A., Lechien, J.R., Ianella, G., Vicini, C., Cammaroto, G., Cannivicci, A., Magliulo, G., Pace, A., Cocuzza, S., Di Luca, M., Stilo, G., Di Mauro, P., Bianco, M.R., Murabito, P., Bannò, V., & La Mantia, I. (2021). Neurocognitive performance improvement after obstructive sleep apnea treatment: State of the art. *Behavioral Sciences*, 11, Article 180.
27. Edinger, J.D., Arnedt, J.T., Berstisch, S.M., Carney, C.E., Harrington, J.J., Lichstein, K.L., Sateia, M.J., Troxel, W.M., Zhou, E.S., Kazmi, U., Heald, J.L., & Martin, J.L. (2021). Behavioral and psychological treatments for chronic insomnia disorder in adults: An American Academy of Sleep Medicine clinical practice guideline. *Journal of Clinical Sleep Medicine*, 17(2), 255–262.
28. Skobic, I., Pezza, M., Howe, G., & Haynes, P.L. (2024). Examining insomnia disorder and stress generation among individuals who have experienced involuntary job loss. *Journal of Psychosomatic Research*, 117, Article 111585.

8 Substance Use

During the detailed interviews we conduct as part of neuropsychological evaluations, we routinely ask patients questions about most aspects of their lives, including health habits like nutrition, sleep, and substance use. Substance use is one of the domains where the range of people's experiences is the most dramatic. We never know what to expect. Some people have never used a single substance recreationally, while others casually report having eight alcoholic drinks a day or smoking marijuana from the moment they wake up until they go to sleep (or they come back from their mid-morning break smelling of it).

This is true across the lifespan. Frequently, my young students assume our older patients do not use substances, but they quickly discover, for example, that many older adults who smoked marijuana in their youth and stopped in middle age are returning to it in retirement, especially given its widening legalization for medical and recreational purposes. Many older adults also increase their alcohol use after they stop working. A saying I have heard a few times from patients living in retirement communities nestled in the stunning foothills of the Santa Catalina mountains goes, "When the mountains turn pink, it's time to drink."

A cocktail with friends at sunset might sound like a harmless habit. A cannabis edible at night might help a patient with chronic pain get much-needed sleep so that they actually wake up refreshed. But what if drinks start at 11 a.m. after a round of golf and continue through the afternoon? When and how does substance use drain our cognitive resources and decrease our functioning?

* * *

In the U.S., more than half of adults use alcohol in any given month, and almost 1 in 4 binge-drink.[1]

Marijuana use varies by age, from over a third of adults 18 to 25 to about 15 percent of those older than 25. About 5 percent of adults misuse a prescription medication (e.g., stimulant, sedative, or pain medications), meaning they use it in a way not directed by a physician: They use more than

DOI: 10.4324/9781003409311-11

prescribed, more often than prescribed, or without a prescription. The most commonly misused medication is pain medication, primarily opioids, misused by 3 percent of adults. If that sounds low, notice that it comes out to over 8 million people in the U.S.

A quarter of young adults 25 and younger, and about 15 percent of those over 25, actually meet criteria for a substance use disorder, meaning their substance use is causing then distress or impairment in their daily functioning, ranging from health problems to inability to fulfill their obligations. However, less than 2 percent of adults receive substance use treatment, and 98 percent of adults with a substance use disorder do not believe they need treatment.

Habitual substance use, with or without the presence of a substance use disorder, can disrupt cognition through multiple mechanisms, including imbalances in neurotransmitters like dopamine, serotonin, and acetylcholine, which, as we have already mentioned, are critical for emotional and cognitive functioning.[2] However, as with other chronic psychological and medical conditions, the cognitive costs of substance use are often omitted from the discourse about substances. This chapter addresses that topic.

Substances and Health

There are many good reasons to cut back on or abstain from alcohol. Alcohol is one of the leading causes of preventable deaths. In the U.S., it contributes to more than 140,000 deaths each year, more than half of those from injuries sustained while intoxicated.[3] Alcohol has been linked to approximately 230 diseases, including infectious diseases, digestive conditions (like acid reflux and gastrointestinal inflammation and bleeding), liver disease, pancreatitis, weakened immune functioning, hematological abnormalities, bone and muscle disease (including increased risk for bone fractures and gout), peripheral neuropathies, endocrine disruptions, lung disease, depression, anxiety, and cardiovascular conditions including hypertension, arrhythmias, cardiomyopathy, atrial fibrillation, heart attacks, and stroke.[2], [3], [4], [5] In people with chronic pain conditions, heavy drinking can exacerbate pain during withdrawal and cause increased pain sensitivity.

Alcohol is also a carcinogen classified in the same risk group as tobacco and asbestos, linked to increased risk of multiple cancer types even at low levels of use.[6] Alcohol use decreases adherence to medical regimens and is related to worse outcomes in chronic conditions including HIV. It is also associated with risky behaviors and increases the risk of motor vehicle accidents, intentional and unintentional injuries, and domestic violence. After excessive drinking, alcohol withdrawal can cause seizures and other life-threatening complications.

You might have heard that moderate drinking can be beneficial for our health. For years, studies showed that those with low to moderate alcohol consumption had lower rates of cardiovascular disease and mortality than those who did not drink at all and those who drink heavily.[5] However, recent research has shown that those findings were due to methodological issues. In many of those studies, light and moderate drinkers were healthier than abstainers, many of whom never drank or stopped drinking because of health problems, or were former drinkers who had to quit because of alcohol abuse. In other words, healthy people were able to drink moderately and lived longer; they were not healthier and lived longer *because* they drank. Newer studies that adjust for those biases have found that moderate drinking is not associated with better health or reduced mortality.[7] In fact, the World Heart Federation has concluded that the message that alcohol can prolong life by reducing the risk of coronary heart disease is a "myth."[4]

In contrast, relatively few adverse health effects have been documented for recreational cannabis use. As with alcohol, marijuana use increases the risk of motor vehicle accidents.[8] Regular smoking of marijuana is associated with worsened and more frequent episodes of some respiratory problems, like bronchitis, and regular cannabis use might increase the risk of cardiovascular disease, heart attack, and stroke.[9] Cannabis use has also been associated with mental health conditions, primarily the development of schizophrenia, and with a mildly increased risk for depressive episodes, suicide thoughts and attempts, and social anxiety.[8] However, these relationships are complex. For example, marijuana use might result in earlier onset of psychosis in those with a genetic predisposition, but it is also the case that those who are in the earliest stages of a psychotic disorder might turn to marijuana in an attempt to manage their mental health symptoms.

How Alcohol and Cannabis Affect Cognition

Alcohol

The most prominent acute effects of alcohol are on focused and divided attention, reaction time, working memory, and response inhibition; at higher doses, it also causes alterations in memory and perception.[10] Low to moderate doses of alcohol significantly reduce glucose metabolism in the whole brain, and the frontal lobes might be the cortical areas most sensitive to the acute effects of alcohol. If people are given alcohol while completing cognitive tasks, there is reduced activation in regions critical for cognitive control, working memory, and error monitoring. Our reaction times increase, working memory decreases, we make more errors, we are less likely to detect those errors, and our decision-making tends to be riskier.

It is easy to see how, even when used in moderation, alcohol can result in cognitive lapses. Imagine you are wrapping up your day after having a couple of drinks. If you are packing your work bag for the next day, you might forget things you need, or you might have to double-check it the next morning to make sure you got everything. If you and your partner make plans about chores and errands, or you read an email about a school event, your recall of what you discussed, agreed upon, and read might be incomplete or inaccurate. Your planning and organization will be less efficient than it would have been without alcohol, or you might forgo any kind of planning or preparation for the next day and then be more rushed in the morning.

As we mentioned in Chapter 7, alcohol also disrupts sleep. Especially when sleep difficulties are due to stress or anxiety, many of us turn to alcohol as a way to induce relaxation and to help us fall asleep. However, alcohol decreases total sleep time and reduces the quality of the sleep we do get, rendering it less restful.[2] Excessive alcohol use can also decrease alertness and dampen cognitive functioning the next day due to the mental fogginess, poor mood, and malaise of even a mild hangover. It also makes it less likely that we will exercise, meditate, or engage in other health behaviors that optimize our cognitive functioning.

In the long term, heavy alcohol use is toxic to the brain, and is associated with decreased brain volume in areas including frontal regions and the hippocampi, loss of white matter, poorer gray and white matter integrity, and disrupted communication between neurons.[11] Alcohol can cause cognitive impairment that is severe enough to meet diagnostic criteria for a major neurocognitive disorder, typically characterized by executive and memory impairments. While some recovery can be seen with abstinence, irreversible and sometimes profound cognitive impairment can occur with severe and prolonged alcohol abuse. Older adults in particular are at increased risk of dementia with sustained, heavy alcohol abuse, and they show reduced capacity for recovery in response to abstinence.

As with cardiovascular health, many studies have reported that moderate drinkers have better cognitive function and decline slower over time than non-drinkers, which led many to claim a "protective" role for moderate drinking on cognition.[5] As we discussed above, more recent research has challenged these findings due to pre-existing health differences between moderate drinkers and non-drinkers, and other studies have shown that even low to moderate alcohol use is associated with an increased risk of cognitive decline and dementia.

Cannabis

In general, cannabis intoxication impairs attention, memory, and executive functions including planning and problem-solving.[8] However, habituation might play a role in these acute effects: Chronic marijuana smokers can actually show some improvement in attention and processing speed after they consume their usual dose.[12] Over time, regular recreational cannabis users experience modest declines in executive functions and in memory (particularly memory for verbal information), especially with moderate and heavy use.[13] Similarly, as a group, long-term regular users of cannabis perform lower than non-users on measures of executive functions, memory, and decision-making.[14] Consistent with this, structural and functional differences have been documented in prefrontal and limbic brain regions between marijuana users and non-users.

Many variables make a difference in the cognitive consequences of regular cannabis use, including how young the person was when they started using it, how long they have been using it, how much they regularly use, tetrahydrocannabinol (THC) concentration, whether the person's pattern of use actually constitutes a cannabis use disorder, whether medical or mental health conditions are present, and even genetic differences.[13] For example, cognitive effects are more pronounced for those who start using during the developmental period, for those who use cannabis more frequently (especially daily), and for those consuming cannabis with higher levels of THC. Fitness also makes a difference: Cannabis users with better aerobic fitness show greater brain volume in multiple brain regions and better neuropsychological performance than those with lower fitness levels. And there might also be a gender effect: Male users might be more vulnerable to neuropsychological deficits overall and might show greater deficits in memory, while female users might show greater changes in attention.

Overall, cognitive changes with regular, recreational cannabis use appear to be small to moderate, do not always indicate the presence of impairment (as with other conditions, there can be some decline, but the person's performance can remain within normal limits), and the declines in cognitive functioning seem to disappear after about a month of abstinence.[13] However, it is important to keep in mind that most studies have been conducted on adolescents and young adults, and it is unclear how cannabis effects interact with the effects of aging.

Medical marijuana use deserves separate mention. People who use cannabis for treatment of chronic pain, post-traumatic stress disorder, or sleep disorders do not show memory decline and can actually show improvement in executive function over time.[15] One important difference seems to be that medical cannabis tends to have lower THC content and higher levels of cannabidiol (CBD). While higher levels of THC have been linked to cognitive decline, CBD seems to mitigate the negative effects of THC on cognition. Moreover, in medical marijuana users, improvements in cognitive performance are related to improvements in the symptoms of the underlying condition (like decreased pain and improved sleep) and to decreased levels of inflammation, and it is CBD content that is associated with these clinical improvements. In other words, medical marijuana might improve cognition thanks to CBD-induced improvement in clinical symptoms.

What To Do About It

Substance use is pervasive in our society. Substance misuse is often normalized and its consequences minimized. This can make it difficult to adequately assess what healthy substance use looks like. Many factors will determine what is right for you, but the information below can be helpful if you want to examine your substance use habits. Remember that just because substances are not causing problems in your life, it does not mean they are not affecting your cognitive functioning and contributing to your lapses.

1. Explore how much is too much for you. First, let's look at some general definitions:

- The often-cited Dietary Guidelines by the U.S. Department of Health and Human Services define *drinking in moderation* as no more than two drinks in a day for men, and no more than one drink in a day for women, on days when alcohol is consumed.[17]
- The National Institute on Alcohol Abuse and Alcoholism (NIAAA) defines *binge drinking* as drinking five or more drinks within a two-hour period for men and four or more drinks within a two-hour period for women.[3]
- The Substance Abuse and Mental Health Services Administration (SAMHSA) defines *heavy alcohol use* as binge drinking on five or more days in a month, based on the thresholds above, while the NIAAA defines *heavy drinking* as consuming five or more drinks on any day or 15 or more drinks per week for men, and consuming four or more drinks on any day or eight or more drinks per week for women.[1], [3]
- The NIAAA defines *high-intensity drinking* as consuming alcohol at levels that are two or more times the gender-specific binge drinking thresholds, which translates into ten or more drinks for men and eight or more drinks for women.[3] People who drink at this level are 70 times more likely to have an alcohol-related emergency room visit. Those who drink three times the threshold are 93 times more likely to have an alcohol-related emergency room visit.

There are no equivalent definitions for cannabis use. While it is possible to determine a *standard alcoholic drink* (defined in the U.S. as a drink containing 14 grams of pure alcohol) to estimate blood alcohol content, no such calculations are available for cannabis, since its effects vary depending on factors including composition (e.g., proportions of THC and CBD) and the method of use (e.g., smoked, inhaled, or ingested in edibles).

Ultimately, however, there is no magic number. What is healthy for you will depend on your cultural context, life circumstances, health, medications you are taking, and other considerations. For example, even one alcoholic drink can be a high risk for an older adult who is physically frail, has an unsteady gait and poor balance, experiences episodes of lightheadedness or dizziness, lives alone in a two-story house, and is on anticoagulant medication that could make a head injury from a fall catastrophic due to the increased risk of a brain bleed.

2. Dig deeper and seek help if needed. Whether your substance use is healthy will not just depend on the amount you consume, but why and how you do it. Why do you use a particular substance regularly? Because it has become a mindless habit? To relax? To change your mood? To control a medical symptom such as pain or insomnia? Because you are bored or it is the main way you socialize? Do you use the substance to feel good, or to stop yourself from feeling bad? Is it working?

Here are other questions that might be helpful:

- Do you ever need to use the substance in the morning to "get going"?
- Do you feel guilt or remorse after using the substance?
- Do you black out or not remember what happened while you were using the substance?
- Are you not doing things that matter to you because of your substance use?
- Has anybody expressed concern about your substance use?[17]

Another resource is the American Psychiatric Association's *Diagnostic and Statistical Manual of Mental Disorders* (DSM) definition of a *substance use disorder*: a problematic pattern of use of a substance (like alcohol, cannabis, opioids, sedatives like benzodiazepines, or stimulants like cocaine) that leads to impairment or distress.[18] While the diagnostic criteria are complex, there might be concern for a substance use disorder if you answer yes to two or more of the following:

- Do you end up using more than you intended to, or for longer periods of time than you intended to?
- Do you want to cut down or control your use, or have you tried (and failed) to cut down or control your use?
- Do you spend quite a bit of time obtaining, using, or recovering from using the substance?
- Do you experience cravings?
- Have you been failing to fulfill important obligations at work, school, or home, because of your use?
- Are you continuing to use despite social or interpersonal problems (e.g., conflict in your relationships) related to your substance use?
- Have you given up or reduced important social, work, or recreational activities because of your use?
- Have you used the substance in hazardous situations (like driving while under its influence)?
- Do you continue to use the substance even though it causes or worsens physical or psychological problems?
- Do you experience *tolerance*, meaning you experience milder effects if you continue to use the same amount, or you need to use more to achieve your desired level of intoxication?

- Do you experience *withdrawal* symptoms, or use the same or another substance to avoid experiencing them?

One more thought for you to consider: The core of *addictive*—habitual, compulsive, and excessive—substance use is a cycling pattern that goes from substance use to negative affect experienced during withdrawal, followed by intense preoccupation with and anticipation of the next time the substance can be used, and restarting the cycle when the substance is used again.[2], [18] If your pattern of substance use fits the definitions above, if you find yourself preoccupied with substances—looking at the clock and calculating how long until you can finally have a drink, smoke a joint, or pop an edible—or if you realize that you use substances as way to cope with unpleasant feelings or unacceptable life circumstances, consider seeking professional help. We should not feel that we *need* substances to go through life.

You can start by talking to your primary care provider, or seek a substance use specialist (see *Resources* below). There are a wide range of interventions that you might benefit from, including support groups, individual counseling, medication, and residential treatment. A specialist can help you determine the issues underlying your substance use and develop a treatment plan that best suits your situation and preferences.

3. *Tend to the garden.* When trying to develop more adaptive substance use habits, relying on healthy habits like physical activity and adequate sleep will be critical, among other reasons, because they optimize the functioning of executive networks that will allow us to monitor and regulate our substance use and resist the urges to over-use. We also need to prioritize finding ways to induce positive emotions. Most recreational substances cause acute surges in dopamine in our brain's reward circuits, and they can increase metabolism in limbic regions involved in reward processing.[10] In other words, substances prime our brain to seek rewards and pleasure, while suppressing top-down cognitive control. If we have been using large amounts of a substance for some time, our dopamine receptors can be less sensitive to natural surges in dopamine from everyday pleasant experiences. [19] Because of this, it will be particularly important to deliberately seek pleasant, pleasurable, fun, and rewarding activities.

Resources

- The website for the National Institute on Drug Abuse and Alcoholism (www.niaaa.nih.gov) and the associated site *Rethinking Drinking* (www. rethinkingdrinking.niaaa.nih.gov) have extensive, reliable information on substance use and its health effects, and resources to examine your alcohol use. There is also a downloadable document, *Treatment for Alcohol Problems: Finding and Getting Help*, with information, advice, and a list of resources.

- The National Cancer Institute (www.cancer.gov) has information on the relationship between alcohol and cancer risk.
- The Substance Abuse and Mental Health Services Administration website (www.samhsa.gov) has educational resources on marijuana use and its effects.
- The National Institute on Drug Abuse (www.nida.nih.gov), the Centers for Disease Control and Prevention (www.cdc.gov), and the American Psychiatric Association (www.psychiatry.org) have information on opioid misuse and its treatment.
- To locate a psychologist with expertise in substance use, you can use the American Psychological Association's Psychologist Locator (https://locator.apa.org) or the *Find a Psychologist* feature of the website for the National Register of Health Service Psychologists (www.findapsychologist.org).

References

1. Substance Abuse and Mental Health Services Administration. (2022). Key substance use and mental health indicators in the United States: Results from the 2021 National Survey on Drug Use and Health. Center for Behavioral Health Statistics and Quality, Substance Abuse and Mental Health Services Administration. www.samhsa.gov/data/report/2021-nsduh-annual-national-report.
2. Yang, W., Singla, R., Maheshwari, O., Fontaine, C.J., & Gil-Mohapel, J. (2022). Alcohol use disorder: Neurobiology and therapeutics. *Biomedicines*, 10, Article 1192.
3. National Institute on Alcohol Abuse and Alcoholism. (2024, February 27). The Healthcare Professional's Core Resource of Alcohol. Knowledge. Impacts. Strategies. www.niaaa.nih.gov/health-professionals-communities/core-resource-on-alcohol/medical-complications-common-alcohol-related-concerns.
4. World Heart Federation. (2022). The impact of alcohol consumption on cardiovascular health: Myths and measures. A World Heart Federation policy brief. https://world-heart-federation.org/news/no-amount-of-alcohol-is-good-for-the-heart-says-world-heart-federation.
5. Zhang, R., Shen, L., Miles, T., Shen, Y., Cordero, J., Qi, Y., Liang, L., & Li, C. (2020). Association of low to moderate alcohol drinking with cognitive functions from middle to older age among US adults. *JAMA Network Open*, 3(6), Article e207922.
6. Anderson, B.O., Berdzuli, N., Ilbawi, A., Kestel, D., Kluge, H.P., Krech, R., Mikkelsen, B., Neufeld, M., Poznyak, V., Rekve, D., Slama, S., Tello, J., & Ferreira-Borges, C. (2023). Health and cancer risks associated with low levels of alcohol consumption. *The Lancet Public Health*, 8(1), e6–e7.
7. Zhao, J., Stockwell, T., Naimi, T., Churchill, S., Clay, J., & Sherk, A. (2023). Association between daily alcohol intake and risk of all-cause mortality: A systematic review and meta-analyses. *JAMA Network Open*, 6(3), Article e236185.
8. National Academies of Sciences, Engineering, and Medicine. (2017). *The health effects of cannabis and cannabinoids: The current state of evidence and recommendations for research.* The National Academies Press.

9. Jeffers, A.M., Glantz, S., Byers, A.L., & Keyhani, S. (2024). Association of cannabis use with cardiovascular outcomes among US adults. *Journal of the American Heart Association*, 13, Article e030178.

10. Bjork, J.M., & Gilman, J.M. (2014). The effects of acute alcohol administration on the human brain: Insights from neuroimaging. *Neuropharmacology*, 84, 101–110.

11. Zahr, N.M., & Pfefferbaum, A. (2017). Alcohol's effects on the brain: Neuroimaging results in humans and animal models. *Alcohol Research: Current Reviews*, 3(2), 183–206.

12. Crean, R.D., Crane, N.A., & Mason, B.J. (2011). An evidence based review of acute and long-term effects of cannabis use on executive cognitive functions. *Journal of Addiction Medicine*, 5(1), 1–8.

13. Lisdahl, K.M., Filbey, F.M., & Gruber, S.A. (2021). Introduction to JINS Special Issue: Clarifying the complexities of cannabis and cognition. *Journal of the International Neuropsychological Society*, 27, 515–519.

14. Lovell, M.E., Akhurst, J., Padgett, C., Garry, M.I., & Matthews, A. (2019). Cognitive outcomes associated with long-term, regular, recreational cannabis use in adults: A meta-analysis. *Experimental and Clinical Psychopharmacology*, 28(4), 471–494.

15. Sagar, K.A., Dahlgren, M.K., Lambros, A.M., Smith, R.T., El-Abboud, C., & Gruber, S.A. (2021). An observational, longitudinal study of cognition in medical cannabis patients over the course of 12 months of treatment: Preliminary results. *Journal of the International Neuropsychological Society*, 27, 648–660.

16. United States Department of Agriculture. (2020). Dietary Guidelines for Americans 2020–2025. www.dietaryguidelines.gov.

17. Babor, T.F., Higgins-Biddle, J.C., Saunders, J.B., & Monteiro, M.G. (2001). *AUDIT. The Alcohol Use Disorders Identification Test: Guidelines for use in primary care* (2nd ed.). World Health Organization.

18. American Psychiatric Association. (2022). Substance use disorders. In *Diagnostic and statistical manual of mental disorders* (5th ed., Text Revision). American Psychiatric Association Publishing.

19. Sterling, P. (2020). *What is health? Allostasis and the evolution of human design*. MIT Press.

9 The Aging Brain

On February 8, 2024, special counsel Robert K. Hur announced the decision to not file criminal charges against U.S. President Joe Biden for mishandling classified documents. The report described the president as a "well-meaning, elderly man with a poor memory" and with "diminished faculties in advancing age." Mr. Hur noted that "Mr. Biden's memory was significantly limited," that he appeared confused about when his vice president term ended, and that he could not remember "even within several years" when his son died.[1], [2]

Articles discussing the report were quick to add that, earlier that week, President Biden had twice referred to a conversation he had with former German chancellor Helmut Kohl in 2021, despite the fact that Mr. Kohl died in 2017.[2] In the press conference where he addressed the special counsel's comments, he referred to the president of Egypt, Abdel Fattah El-Sisi, as the president of Mexico.[2] Dismissals from his supporters came fast, pointing out that former President Donald Trump had similarly confused the leaders of Hungary and Turkey, had said the country was on the verge of World War II, had said he defeated Barack Obama instead of Hillary Clinton, and had referred to Nikki Haley as Nancy Pelosi.[3] A predictable headlines war ensued:

> Is Biden an out-to-lunch president?[4]
> Trump said he's running against Obama. Stop downplaying his memory lapses.[5]
> How old is too told to be president?[6]

* * *

Unlike the issues covered in previous chapters, aging is one cause of cognitive changes that we will all have to face—if we are lucky. And our chances of enjoying older age are getting better: In 2019, the average global life expectancy reached 73 years, an increase of almost nine years since 1990 (although with a quite astonishing range, with life expectancies around the world ranging from the mid-50s to the mid-80s).[7] By 2050, average

DOI: 10.4324/9781003409311-12

global life expectancy is expected to be around 77, and the number of persons 65 and older will be more than twice the number of children under five.

While this rapid aging of the global population reflects a success—decreased mortality—it also brings significant public health challenges, because the number of people living with age-associated conditions like dementia will increase tremendously over the next few decades: The World Health Organization (WHO) estimates that there are currently over 55 million people living with dementia around the world, and this number is expected to triple by 2050.[8] The urgency of promoting healthy aging is reflected in the United Nations' 2020 adoption of a resolution proclaiming 2021–2030 the Decade of Healthy Ageing, a global collaboration, led by the WHO, with the goal to improve the lives of older people by combating ageism, creating age-friendly environments, providing comprehensive medical and mental healthcare, and ensuring long-term care for those who need it.[9]

The Secret Life of Brains Is Complex

Before we review what cognitive changes are normal with aging, it is important to mention a few points that highlight the complexity of the cognitive aging process. First, there is great variability in the cognitive changes people experience with aging.[10] In fact, there is evidence that inter-individual variability—meaning, in this case, differences in cognitive performance among individuals of the same age group—increases with age. In other words, there is a wide range of what is considered "normal" cognitive performance in older age. The same is true for age-related brain changes: Brains age differently, at different times and at a different pace, and there is a wide range of what a "normal" older brain looks like.[11]

Second, many of the changes that we consider typical of aging start earlier in life. Throughout development, from the prenatal to older age stages, brains change in different ways and at different rates—for example, explosively developing connections, then steadily pruning them.[12] Mild declines in some aspects of cognitive functioning and brain volume begin as early as our 30s, with the rate of decline accelerating in our 60s.[10], [11], [12] So it is important to see aging in this developmental context, and to think of adapting to the changing needs of our brains as a lifelong process.

Third, not only do our brains age differently, but we differ significantly in our ability to endure the changes that occur with aging and even those that occur with brain disease. Many older adults walk around with substantial neuropathology (e.g., the abnormal accumulation of proteins characteristic of Alzheimer's disease) without showing any cognitive symptoms.[13] Why is this? One answer is *cognitive reserve*: Individuals with higher levels of educational and occupational achievement, for example, seem to be able to withstand higher levels of neuropathology without developing cognitive symptoms, while people with lower educational or occupational levels are more likely to show cognitive impairment with the same (or even lower) levels of pathology in their brain. This suggests that decades of cognitively stimulating activity might give brains a reserve of resources that allows them to continue to function at a normal level despite the presence of disease.[14] In other words, brains that have lived cognitively rich lives develop symptoms only when the disease is more advanced.

Fourth, socioeconomic variables are a major determinant of health in general and have a powerful influence over how our brains age. People of lower socioeconomic status (SES) are at higher risk for premature mortality, serious health conditions, and dementia.[15] Older adults of lower SES, for example, show greater declines in physical, sensory, cognitive, emotional, and social functioning than older adults of higher SES, independently of their baseline health status. Disparities in brain health in individuals from ethnoracial minorities have been documented starting in midlife, with Black adults showing signs of brain aging beginning in midlife and Hispanic and White adults displaying signs of brain aging later in life.[16] Black older adults are also twice as likely to have dementia and Hispanic older adults one and a half times more likely to have dementia as White older adults. [17] These findings suggest that the cumulative impact of socioeconomic adversity can lead to accelerated brain aging and neurodegeneration.

Finally, many physical and sensory changes common in older age—like decreased visual acuity and hearing sensitivity—can mimic, exacerbate, or mask cognitive changes.[18] Older adults living with uncorrected hearing loss, for example, might seem inattentive, forgetful, and confused to others, because they miss or mishear things that are said. On the other hand, well-meaning family members might misattribute an older adult's cognitive impairment to the fact that they cannot hear or see well.

So What Is "Normal" Aging?

Age-related brain *atrophy* refers to a decrease in brain volume with age—yes, our brains shrink. As neurons dwindle and the connections among them diminish, there is cortical volume loss and cortical thinning, most prominently in prefrontal and temporal areas important for executive and memory functions. There are decreases in white matter volume, also more pronounced in frontal regions, and an increase in white matter lesions. There are changes in neurotransmitters—for example, with loss of dopamine receptors, part of the dopamine system that is critical to cognition and attention regulation in particular.[19]

Consistent with these brain changes, we experience declines in several cognitive domains.[19], [20], [21], [22] Slower processing speed is a key cognitive change with aging. On commonly used tests of processing speed, the average performance for an 80-year-old would be in the impaired range for a 20-year-old, and the average performance for a 20-year-old can only be attained by less than the top 1 percent of 80-year-olds.[23]

In the memory domain, we experience declines in episodic memory. Our overall memory capacity diminishes, and we remember less information than when we were younger. On commonly used memory tests, the average 80-year-old recalls two-thirds the amount of information an average 20-year old recalls.[24] We still benefit from repetition and show a *learning curve*, meaning we recall more information each time the information is repeated, but again the average 80-year-old learns about three-quarters the amount of information an average 20-year old learns with repetition. But while we learn less information than before, in healthy aging we remain able to retain the information we learn.

There is noticeable decline in our ability to recall details, while gist memory remains largely preserved. For example, a healthy older adult will be able to remember that they attended a large family dinner a couple of weeks ago in a new restaurant on the north side of town, but they might not remember the name of the restaurant, what appetizer they ordered, or who sat next to whom.

Our *source memory* declines; this refers to when, where, and from whom we learned information. The same healthy older adult might remember learning that evening that Bob and Sheila are getting divorced, but they might not recall who said it. This can make older adults vulnerable to misinformation, since they might recall hearing about a new medication, for example, and not recall if their physician told them about it or if they heard it in some ad making unsupported claims. Prospective memory, the ability to remember do something in the future (like return that expensive item we unwisely purchased before the deadline to get a refund) also declines, which is why a common complaint from older adults is "I have to write *everything* down now."

Less effortful aspects of memory, like feelings of familiarity, tend to remain preserved in healthy aging. This is why we are often able to recognize information we cannot recall. For example, we might run into Bob and Sheila at the grocery store and not be able to recall Sheila's name, but if someone says her name is Sheila, we will say, "Of course! Sheila!" The information has not been forgotten; we just have difficulty retrieving it, especially retrieving it quickly. This is in contrast with a patient with true amnesia, as in Alzheimer's disease, where the information has actually been forgotten, so if someone says, "That's Sheila," the name does not sound familiar to them anymore (even if they still say, "Oh, right...").

While basic attention remains largely preserved, there are widespread changes in executive functions. Our working memory capacity is reduced, so we are more likely to "forget" information that we were trying to keep in mind for a brief period of time (this is why it becomes so common to walk into a room and not remember why). Slower processing speed and reduced working memory account for many of our difficulties in daily life: We slow down but the world does not, and we miss some of what is said and happens around us because we cannot process as much information or as fast as before.

Cognitive inhibition diminishes, and we become more distractible because we have difficulty suppressing irrelevant stimuli around us, and even our own irrelevant thoughts—this might be one reason we become more tangential. There are also decrements in divided attention and set-shifting, which is why multitasking becomes so difficult. (My 20-year-old can play videogames while talking to his friends on the headset and checking social media on his phone, but I have to turn the music off and tell everyone in the car to stop talking if I have to parallel park.) This diminished ability to shift is one of the most pronounced changes that occurs as we age: On commonly used tests of switching, a performance that is average for an 80-year-old is only attained by the bottom 0.1 percent of 20-year-olds. Even a performance that is average for a 50-year-old would place a 20-year-old in the bottom 2 percent of their age group.[25]

In general, automatic processing remains largely preserved, while effortful processing—when we have to stop and deliberately *think*—becomes more difficult. This is why we become more cognitively *in*flexible as we age and why we become very attached to our routines, because anything requiring cognitive control, adaptation, and flexibility becomes more effortful than when we were younger. Given all these executive changes, it is not surprising that we struggle more with novel problem-solving, like figuring out how to use new technology. We can do it, but it takes more time and energy.

In the language domain, a common age-related change is reduced efficiency when retrieving words; this is why word-finding difficulties ("The… the…the thing" while gesticulating wildly) and "tip of the tongue" lapses are common. Visuospatial abilities decrease too, and it takes us longer, for example, to figure out routes and find our way around unfamiliar places.

Overall, cognitive changes with age tend to decrease the efficiency of our cognitive performance. Like the other conditions reviewed in this section, with aging, things that used to be automatic now require more effortful control, spending more cognitive resources and decreasing our capacity to pay attention to other things. Consistent with this idea that cognitive processing becomes less efficient and more effortful as we age, there is evidence that more widespread brain activation during cognitive tasks (e.g., activation in both hemispheres instead of one, and increased activation in frontal areas) is associated with better performance in older adults. The older brain has to engage in additional, top-down compensatory processing, requiring activation of more widely distributed areas of the brain to maintain cognitive performance.

* * *

To go back to our presidents, misspeaking and retrieving the wrong term or name, especially in a rushed or stressful situation, is normal in older age. For example, if we were recently thinking about Nancy, we might call Nikki

"Nancy" due to a failure to inhibit "Nancy." Similarly, if we were recently talking about Mexico, we might say "Mexico" when we meant "Egypt." What would be a reason for concern is if someone very familiar with both Nikki and Nancy seems to have forgotten which is which, or if someone very familiar with the president of Egypt seems to forget which country he is a president of, especially if reminders do not help. Such loss of knowledge is not something we can easily attribute to aging. The Appendix provides information about when cognitive changes are reason for concern warranting a formal evaluation.

The Wisdom of Aging Brains

It is very important to point out that cognitive aging is not all about losses and decline. Vocabulary continues to increase across the life span. Our consolidated semantic knowledge (our stored general knowledge) remains preserved, and we are able to continue to learn, although it might take more time, practice, or effort. Procedural knowledge is similarly preserved: We do not lose our cognitive ability to engage in activities we are proficient at—playing the piano, knitting, or replacing an alternator—even if we do experience limitations from physical problems.

In the emotional domain, older adults report higher levels of well-being despite the physical, cognitive, and social losses that aging can bring.[26] Older adults report lower levels of psychopathology, better control over their emotions, more positive emotional experiences, and higher levels of empathy, gratitude, and forgiveness, with lower rates of worry, anger, and sadness. As we age, the perception of our remaining time as limited results in a shift to prioritizing emotionally meaningful goals, like time with loved ones, over exploration goals, like visiting new and exotic places. We prune our social networks, focusing on those we are emotionally close to.

There is even a *positivity effect* in cognitive processing: While for most of our life we have an attentional bias towards negative information, which can be relevant for our survival, the bias becomes positive with age.[27] Healthy older adults spontaneously tend to focus their attention on and remember positive information more than negative information. There is also evidence that, when under stress, older adults' memory and executive functioning might be affected less than those of younger adults, and that stress might in fact shift older adults' decision-making towards more careful, less risk-averse strategies.[28]

These cognitive and emotional strengths can result in what we call *wisdom*, a complex quality composed of general knowledge of life and social decision-making, effective emotional regulation, prosocial behaviors like compassion and empathy, insight or self-reflection, acceptance of different value systems, and decisiveness.[29] This is why older adults can excel at complex problem-solving that requires knowledge, expertise, and the consideration of multiple perspectives, when it does not occur under time pressure. This is also why multigenerational teams in work settings benefit from a combination of the quick thinking, creative, rapidly adapting, technologically savvy younger brains, and the expert knowledge, practical experience, nuance, and cautious thinking of older brains. Wisdom does not come from longevity alone, but there is no wisdom without life experience.

What To Do About It

As we age, we might need to change how we do things to avoid mistakes and lapses. Our performance can improve in meaningful ways with the supports and strategies we will review in Part III. In addition, here are a few things to work on in the longer term.

1. Do not buy into negative stereotypes of aging. A large body of research has documented the harmful effects of holding negative views of aging.[30], [31], [32], [33], [34], [35], [36], [37] Adults who hold more negative stereotypes of aging are more likely to experience a cardiovascular event like a heart attack or stroke and to experience them earlier in life, are more likely to develop chronic pain, lose hippocampal volume faster, and have significantly higher levels of Alzheimer's pathology than those with more positive views of aging. Adults with more positive views of aging have better overall physical and mental health; if they are veterans, they are less likely to experience depression, anxiety, PTSD, and suicidal thoughts. They are more likely to engage in healthy behaviors like physical activity, are less likely to engage in harmful behaviors such as smoking, recover better after a heart attack, show better cognitive performance, are less likely to experience cognitive decline or develop dementia (even those with a genetic risk factor), and live anywhere from two and a half to seven and a half years longer.

How can beliefs be so powerful? We probably start assimilating negative stereotypes of aging from society, sometimes unconsciously, in childhood. When these beliefs become self-relevant later in life, they can impact our health through negative emotions like hopelessness, cognitive biases like blaming all physical changes on aging, and behaviors like physical inactivity.[38] If we believe that older age is inevitably a time of physical frailty, cognitive impairment, and dependence, we are less likely to take active steps to preserve our health and more likely to live chronically stressed or even depressed.

Even if we do not personally hold negative aging stereotypes, ageism hurts us. When older adults are exposed to negative aging stereotypes like the words "frail" and "confused," even when the words are presented too fast for them to consciously know what they saw, they perform worse on subsequent cognitive and balance tests, show heightened cardiovascular responses to stress-inducing tasks, and are less likely to accept life-prolonging interventions when presented with hypothetical scenarios describing a potentially fatal illness.

So be very deliberate about informing yourself about what healthy aging can look like, seeking age-friendly spaces, avoiding—and speaking up against—negative aging stereotypes, and seeking care from geriatricians and other aging-informed physicians.

2. Seek cognitive stimulation. The concept of *neuroplasticity* refers to the ability of the brain to change in structure (e.g., by developing new connections) or function (e.g., by increasing or decreasing activity in certain areas) in response to stimulation. In other words, neuroplasticity refers to the ability of the brain to learn and adapt. While older adults show less neuroplasticity than younger adults, the aging brain does retain considerable plasticity.[21] For example, older adults who suffer a stroke resulting in loss of brain tissue do show recovery with rehabilitation, indicating that the aging brain is capable of reorganizing itself and regaining function. Plasticity is triggered by challenges, by demands on our cognitive systems. Cognitive activities that are novel and challenging are key to capitalize on our brain's potential.

Does this mean you should spend time (and money) on websites and apps offering "brain games"? No. You can, if you enjoy them, but claims about their benefits are notoriously exaggerated and unsubstantiated (the same is true of the many "brain health" supplements out there, unfortunately). Consistently engaging in varied, complex, novel activities that you find challenging but still doable, so you can experience enjoyable successes with effort, does benefit your brain and cognitive functioning, with bonus points if the activities are also social: You can volunteer as a docent in a museum, sign up for classes to learn a new hobby that does *not* come easily to you, or regularly attend community lectures with friends.

3. Tend to the garden. A healthy older brain can achieve remarkable things, make invaluable social contributions, and support unprecedented levels of well-being. Start by making sure physical and perceptual issues are addressed with, for example, reading glasses, hearing aids, and mobility supports.

It is also extremely important to address mental health issues: Depression in older adults increases the risk for a variety of medical illness, and medical problems increase the risk of late-life depression.[39] Depression symptoms like apathy and withdrawal, especially if they appear for the first time in older age, can also be early signs of a neurocognitive disorder. For all these reasons, prompt assessment of mood changes is critical.

Remaining as physically active as safely possible, cognitively challenged, and socially engaged are the best gifts we can offer our brains as we age; protecting our sleep and ensuring we are rested and alert make all the other habits possible. Neuropsychologists routinely recommend this trifecta of physical, cognitive, and social activity because we know just how powerful it is. While we might not be able to avoid age-associated changes, these habits can help us optimize our functioning. At any point in time, there is a range within which we can possibly function. When depleted, we function at our lower limit, and when healthy and active, we can function at our best.

Moreover, behavioral changes that improve our cardiovascular, overall medical, and mental health are so effective at preventing or delaying cognitive decline that together they could reduce dementia cases by a third.[40] Even if we cannot avoid developing dementia, a significant delay in the onset of symptoms would buy us precious time, and who among us would refuse a chance at more healthy years doing what we love with the ones we love?

Resources

- Two very helpful and accessible books about healthy brain aging, both written by clinical neuropsychologists, are *Keep Your Wits About You: The Science of Brain Maintenance as You Age*, by Vonetta Dotson, and *High-Octane Brain: 5 Science-Based Steps to Sharpen Your Memory and Reduce Your Risk of Alzheimer's* by Michelle Braun.
- The American Geriatrics Society has a website (www.HealthInAging. org) with valuable and reliable information for older adults and their care partners.
- The National Center to Reframe Aging was founded by a group of organizations led by the Gerontological Society of America, with the mission to end ageism. The Center's website (www.reframingaging.org) has many educational brochures and videos openly available (its Learning Center, with more in-depth trainings, requires the creation of a free account).
- The Alzheimer's Association (www.alz.org) publishes a free, downloadable annual report, *Alzheimer's Facts and Figures*, with extensive information on Alzheimer's and other dementias.
- To learn about "Super Agers," older adults whose cognitive abilities remain at levels expected for individuals who are decades younger, visit the University of Southern California's SuperAgers website (www.gero.usc.edu/cga/superagers). You can also learn about the SuperAger Research Initiative at the University of Chicago's Healthy Aging & Alzheimer's Research Care Center at (www.haarc.center.uchicago.edu/superagers).

References

1. Madhani, A., Peoples, S., & Long, C. (2024, February 9). Takeaways from the special counsel's report on Biden's handling of classified documents. *Associated Press*. https://apnews.com/article/biden-classified-documents-age-trump-2024-4791639cc06cc0affee55aba80c7e6b3.
2. Montanaro, D. (2024, February 10). Biden's rough week highlights his biggest vulnerability—one he can't change. *NPR*. www.npr.org/2024/02/10/1230591708/biden-age-special-counsel-report-classified-documents.
3. Shear, M.D. (2024, February 8). Special Counsel's report puts Biden's age and memory in the spotlight. *The New York Times*. www.nytimes.com/2024/02/08/us/politics/biden-special-counsel-report-documents.html.

4. Bergen, P. (2024, February 8). Is Biden an out-to-lunch president? *CNN*. www.cnn. com/2024/02/08/opinions/biden-special-counsel-report-memory-bergen/index.html.

5. White, J.K. (2024, February 20). Trump said he's running against Obama. Stop downplaying his memory lapses. *The Hill*. https://thehill.com/opinion/campaign/4476919-trump-said-hes-running-against-obama-stop-downplaying-his-memory-lapses.

6. Baker, P. (2024, February 10). How old is too old to be president? An uncomfortable question arises again. *The New York Times*. https://www.nytimes.com/2024/02/10/us/politics/biden-trump-age.html.

7. United Nations Department of Economic and Social Affairs, Population Division. (2022). *World Population Prospects 2022: Summary of Results*. UN DESA/POP/2022/TR/NO. 3.

8. GBD 2019 Dementia Forecasting Collaborators. (2022). Estimation of the global prevalence of dementia in 2019 and forecasted prevalence in 2050: An analysis for the Global Burden of Disease Study 2019. *Lancet Public Health*, 7, e105–e125.

9. Decade of Healthy Ageing. (2022). Decade of healthy aging: The platform. www. decadeofhealthyageing.org.

10. LaPlume, A.A., Anderson, N.D., McKetton, L., Levine, B., & Troyer, A.K. (2021). When I'm 64: Age-related variability in over 40,000 online cognitive test takers. *Journals of Gerontology: Psychological Sciences*, 77(1), 104–117.

11. Fujita, S., Mori, S., Onda, K., Hanaoka, S., Nomura, Y., Nakao, T.et al. (2023). Characterization of brain volume changes in aging individuals with normal cognition using serial magnetic resonance imaging. *JAMA Network Open*, 6(6), e2318153.

12. Bethlehem, R.A.I., Seidlitz, J., White, S.R., Vogel, J.W., Anderson, K.M., et al. (2022). Brain charts for the human lifespan. *Nature*, 604, 525–533.

13. Ossenkoppele, R., Binette, A.P., Groot, C., Smith, R., Strandberg, O., Palmqvist, S., Stomrud, E., Tideman, P., Ohlsson, T., Jögi, J., Johnson, K., Sperling, R., Dore, V., Masters, C.L., Rowe, C., Visser, D., van Berckel, B.N.M., van der Flier, W.M., Baker, S. … Hansson, O. (2022). Amyloid and tau PET-positive cognitively unimpaired individuals are at high risk for future cognitive decline. *Nature Medicine*, 28, 2381–2387.

14. Stern, Y. (2021). How can cognitive reserve promote cognitive and neurobehavioral health? *Archives of Clinical Neuropsychology*, 36, 1291–1295.

15. Steptoe, A., & Zaniotto, P. (2020). Lower socioeconomic status and the acceleration of aging: An outcome-wide analysis. *PNAS*, 117(26), 14911–14917.

16. Turney, I.C., Lao, P.J., Renteria, M.A., Igwe, K.C., Berroa, J., Rivera, A., et al. (2022). Brain aging among racially and ethnically diverse middle-aged and older adults. *JAMA Neurology*, 80(1), 73–81.

17. Alzheimer's Association. (2024). Alzheimer's disease facts and figures. *Alzheimer's & Dementia*, 20(5).

18. Cavazzana, A., Röhrborn, A., Garthus-Niegel, S., Larsson, M., Hummel, T., & Croy, I. (2018). Sensory-specific impairment among older people: An investigation using both sensory thresholds and subjective measures across the five sense. *PLOS One*, 13(8), e0202969.

19. Mosti, C.B., Rog, L.A., & Fink, J.W. (2019). Differentiating mild cognitive impairment and cognitive changes of normal aging. In L.D. Ravdin & H.L. Katzen (Eds.). *Handbook on the neuropsychology of aging and dementia* (2nd ed.). Springer.

20. Park, D.C., & Reuter-Lorenz, P. (2009). The adaptive brain: Aging and neurocognitive scaffolding. *Annual Review of Psychology*, 60, 173–196.

21. Park, D.C., & Bischof, G.N. (2013). The aging mind: Neuroplasticity in response to cognitive training. *Dialogues in Clinical Neuroscience*, 15, 109–119.
22. Drag, L.L., & Bielauskas, L.A. (2010). Contemporary review 2009: Cognitive aging. *Journal of Geriatric Psychiatry and Neurology*, 23(2), 75–93.
23. Wechsler, D. (2008). *WAIS-IV administration and scoring manual*. NCS Pearson.
24. Delis, D.C., Kramer, J.H., Kaplan, E., & Ober, B.A. (2017). *California Verbal Learning Test, Third Edition (CVLT-3)*. Pearson.
25. Delis, C.D., Kaplan, E., & Kramer, J.H. (2001). *Delis-Kaplan Executive Function System examiner's manual*. The Psychological Corporation.
26. Carstensen, L.L. (2021). Socioemotional selectivity theory: The role of perceived endings in human motivation. *The Gerontologist*, 61(8), 1188–1196.
27. Reed, A.E., Chan, L., & Mikels, J.A. (2014). Meta-analysis of the age-related positivity effect: Age differences in preferences for positive over negative information. *Psychology and Aging*, 29(1), 1–15.
28. Mikneviciute, G., Ballhausen, N., Rimmele, U., & Kliegel, M. (2022). Does older adults' cognition particularly suffer from stress? A Systematic review of acute stress effects on cognition in older age. *Neuroscience and Biobehavioral Reviews*, 132, 583–602.
29. Jeste, D.V., & Lee, E.E. (2019). Emerging empirical science of wisdom: Definition, measurement, neurobiology, longevity, and interventions. *Harvard Review of Psychiatry*, 27(3), 127–140.
30. Levy, B.R., & Leifheit-Limson, E. (2009). The stereotype-matching effect: Greater influence on functioning when age stereotypes correspond to outcomes. *Psychology and Aging*, 24(1), 230–233.
31. Levy, B.R., Zonderman, A.B., Slade, M.D., & Ferruci, L. (2009). Age stereotypes held earlier in life predict cardiovascular events in later life. *Psychological Science*, 20(3), 296–298.
32. Levy, B.R., Ferruci, L., Zonderman, A.B., Slade, M.D., Troncoso, J., & Resnick, S. M. (2016). A culture-brain link: Negative age stereotypes predict Alzheimer's disease biomarkers. *Psychology and Aging*, 31(1), 82–88.
33. Levy, B.R., Chung, P.H., Slade, M.D., Van Ness, P.H., & Pietrzak, R.H. (2019). Active coping shields against negative aging self-stereotypes contributing to psychiatric conditions. *Social Science & Medicine*, 228, 25–29.
34. Levy, B.R., Slade, M.D., & Lampert, L., (2019). Idealization of youthfulness predicts worse recovery among older individuals. *Psychology and Aging*, 34(2), 202–207.
35. Levy, B.R., Pietrzak, R.H., & Slade, M.D. (2023). Societal impact on older persons' chronic pain: Roles of age stereotypes, age attribution, and age discrimination. *Social Science & Medicine*, 323, Article 115772.
36. Ng, R., Allore, H.G., Monin, J.K., & Levy, B. R. (2016). Retirement as meaningful: Positive retirement stereotypes associated with longevity. *Journal of Social Issues*, 72(1), 69–85.
37. Tully-Wilson, C., Bojack, R., Millear, P.M., Stallman, H.M., Allen, A., & Mason, J. (2021). Self-perceptions of aging: A systematic review of longitudinal studies. *Psychology and Aging*, 36(7), 773–789.
38. Levy, B. (2009). Stereotype embodiment: A psychosocial approach to aging. *Current Directions in Psychological Science*, 18(6), 332–336.
39. Alexopolous, G.S. (2019). Mechanisms and treatment of late-life depression. *Translational Psychiatry*, 9, 188.

40. Yaffe, K., Vittinghoff, E., Dublin, S., Peltz, C.B., Fleckenstein, L.E., Rosenberg, D. E., Barnes, D.E., Balderson, B.H., & Larson, E.B. (2024). Effect of personalized risk-reduction strategies on cognition and dementia risk profile among older adults: The SMARRT randomized clinical trial. *JAMA Internal Medicine*, 184, 54–62.

Part III

Decreasing Cognitive Overload

We have seen how, when our brains are depleted (by chronic stress, medical or psychological conditions, sleep deprivation, or substance use), we experience attention lapses, our working memory capacity is decreased, we have less ability to inhibit distractions, we have a more difficult time shifting our attention, and we process information more slowly than usual. These problems with attention, executive functions, and processing speed can, in turn, cause memory failures. We glitch all around. Ideally, we will be able to follow the recommendations from Part II and actually minimize the burdens that deplete our brain.

Part III addresses the second pillar of a brain-friendly life: Reducing excessive cognitive demands. We will first, in Chapter 10, review characteristics of modern life that contribute to cognitive overload. Then, in Chapters 11 through 21, we will delve into strategies that we can implement right away to decrease the demands on our precious cognitive resources, provide our brain with structures and supports, and help us glitch less in daily life even when our brains remain somewhat cognitively drained.

Of course, not every strategy will be helpful every time. Some days, you will mostly need to slow down, take breaks, and ground yourself; some days you will need to focus on maintaining your alertness and perfecting your time management so you can efficiently get everything done. Some days you will be able to implement many of these strategies; some days you might only be able to practice one of them. Each one is meant to be a tool that you can use depending on your circumstances and needs at any given time.

As you read on, you might find yourself wondering if these strategies can actually make a difference when you are depleted and overloaded. Yes, they can. Even small changes in your days can have quite a significant impact on your functioning. And, after all, you are most likely holding this book because what you have been trying has not been working, so give these a try.

DOI: 10.4324/9781003409311-13

10 Life

One reason we experience cognitive overload is that many of us live in modern environments that impose excessive demands on our brains. Some characteristics of modern life that make it brain-unfriendly include a sense of chronic time pressure, technology that brings the entire world to the palm of our hands, the spreading out of our social support networks, spending our days in systems developed for productivity and efficiency and not well-being or comfort, and the normalization of unrealistic expectations regarding success, parenting, relationships, and wealth. In this chapter, we will review some of these factors that impose excessive cognitive demands and drain our cognitive resources, resulting in cognitive overload and glitching.

Technology and Media

We cannot talk about modern life without talking about the ubiquitous presence of digital technology. There is wide variability in technology use across age groups, but overall, adults in the U.S. spend four and a half hours a day on their mobile phone for reasons other than talking, and over 11 hours a day consuming digital content across all screen types (smartphones, TVs, game consoles, computers, tablets, and others).[1], [2] Forty percent say they are online almost constantly.[3]

These numbers are not an accident: Technology is designed to capture and keep our attention. Tristan Harris, former Google Design Ethicist, and Aza Raskin, a former interface designer, use their insider knowledge to advocate for the ethical use of technology and educate the public on how technology is "strategically designed to compel us to engage."[4] Constant notifications catch our attention and can make us feel we are missing something important if we do not check our apps. Newsfeeds use sensationalized language and shocking images to evoke negative states like uncertainty, anxiety, or outrage, which our brain's natural alarm systems prioritize and have difficulty disengaging from. Features like infinite scrolling, suggested content, and auto-play of videos effectively take away our choice and numb our will, making us fall into a mode of passive, never-ending content consumption.

DOI: 10.4324/9781003409311-14

The problem goes beyond how much time we spend using technology: *How we consume media* has changed dramatically. We used to have a choice between a few shows at a time, on a handful of TV channels whose broadcasts ended at a certain hour. We now have access to endless options for round-the-clock entertainment and news at our fingertips. We used to learn about a foreign war by reading a newspaper article with a few black-and-white pictures. Now, we can access (and at times we are exposed to, without asking for it) videos of devastating tragedies happening everywhere in the world. We can literally see and hear the sights and sounds of destruction, violence, and suffering, sometimes in real time, any time, wherever we are.

And then there is social media. Globally, the average person spends two and a half hours a day on social media,[5] although again there is large variation by age, with the average U.S. teenager spending almost five hours a day on social media apps.[6] Social media introduces additional problems. It encourages social comparison, often against highly manipulated, carefully curated content. It exposes us to unrealistic expectations for our bodies, our family life, and our homes, then inundates us with ads for products to address those made-up inadequacies.

Interestingly, it might not be the time we spend on social media per se that impacts our psychological well-being. Amount of exposure to social media is associated with slightly higher levels of depression and anxiety, but, importantly, it can also be associated with mildly stronger feelings of belonging and connectedness.[7] It is engaging in upward social comparisons on social media, specifically, that has negative effects on our body image, mental health, sense of well-being, life satisfaction, and self-esteem. [8] So *how* we use social media, in addition to *how much*, has important effects on our mental health.

It is not difficult to see that our brains did not evolve to process information like this. As we reviewed in Chapter 1, our brains evolved to keep us alive by anticipating our needs and threats in the environment, and developing flexible behavioral adaptations to satisfy those needs and avoid those threats. We are also social animals whose brains automatically process others' emotional states. Through technology that follows us everywhere, we are now constantly exposed to virtual "threats" and the distress of others. We can see and hear wars, natural disasters, and violence during a quick coffee break, in the playground, and in our bed. We are also exposed to culturally constructed social "threats" to our self-image, self-esteem, and life satisfaction. The "others" that matter are no longer just the few dozen from our tribe, but thousands of strangers online. These constant exposures trigger our stress response and set our brain on alert. We are then supposed to put our phone down and help the kids with their homework, tend to customers, or fall asleep. This requires an exceptional amount of cognitive inhibition and emotional regulation, which is extremely cognitively costly.

There is another major way technology affects our brains. Twenty years ago, people spent, on average, two and a half minutes on any specific content on a screen; in the last few years, the average has been 47 seconds. [9] The amount and pace of media have shortened our attention spans—or at least our attention preferences—which means we are constantly shifting our attention from one content or task to another. In Chapter 15, we will see how set-shifting is stressful, cognitively expensive, and causes us to make more mistakes and take longer to complete tasks. Constant shifting throughout the day, every day, can add significantly to our feelings of overwhelm and is one major reason we glitch so much.

Work

Globally, working hours for the average worker have decreased dramatically. It is difficult, however, to make generalizations based on the "average" worker, because there is wide variability among sectors of society and industry. For example, globally, the average number of hours worked per week is 44 hours, but a third of workers regularly work more than 48 hours per week.[10] There is also large variability among countries: U.S. workers work more hours and have significantly less time off than other developed European countries like the U.K. and Germany, but work fewer hours that those in China, India, and Korea.

Work is commonly identified as a source of stress. Seventy-seven percent of American workers report experiencing work-related stress, and over half report symptoms of burnout like emotional exhaustion, lack of motivation, and a desire to quit.[11] Only a third report that their work culture encourages taking breaks and only 40 percent say it respects their time off. One important contributor to work-related stress is that many aspects of our work lives are outside of our control and determined by external factors like productivity and cost-efficiency for the employer, not employees' health or well-being. Those workers who are satisfied with how much control they have over how, when, and where they work, and those with better balance between their work and home lives are more likely to report better mental health.

Work environments and workdays are not always worker-friendly. The modern office was transformed in the 1960s due to increased need for office space; companies' demand for cheap and space-saving furniture gave us the cubicle.[12] With the advent of personal computers, workers started spending more time sedentary at their desks and in front of screens. Many workers have little control over their schedule, including when to take breaks or eat, and activities are often scheduled back-to-back, resulting in the need for rapid shifting, with no room for mental transitions. While recently there has been more emphasis on well-being at work, such efforts often reveal that the ultimate motivation is still increased productivity.

Against this backdrop, the Covid pandemic upended office life and increased the reliance on remote work. Currently, one-third of American workers do some or all of their work from home.[13] Most workers are happy with this change: Over 80 percent of them want to retain some form of remote work, and over a third would like to work remotely full-time. While remote work has consistently been found to increase productivity and job satisfaction, it also introduces problems, most notably "technology exhaustion" (due to its reliance on virtual work spaces) and the blurring of the boundaries between work and home.[14] Again, our brains find themselves constantly shifting: We turn our camera off during a meeting to see why the dog won't stop barking, we repeatedly stop in the middle of a budget calculation to answer our child's questions, and we write work emails from the couch while the plumber works on the kitchen sink. This shifting leads to increased stress, fatigue, and errors.

Chronic Time Pressure

One of the most pervasive experiences in modern life is the feeling that we simply do not have enough time. Many of us feel we do not have time to schedule medical appointments and home repairs, that technological time-saving devices in fact make time management more complicated, or that the hours between work and bedtime fly by, filled with an unreasonable number of activities and chores.

The world has, indeed, gotten faster, but this has been going on for a very long time. The increasing speed of our days—and the psychological sense of chaos it can bring along—has long been recognized by social theorists, though it is largely seen as inseparable from social forces like industrialization, individualism, and urbanism—an inevitable part of what we call modern life. In recent decades, the world seems to have accelerated at a faster pace, leading to what has been described as a culture of immediacy, with expectations of next-day deliveries and instant access to the digital content we crave.[15]

The content we consume has itself accelerated. The pace of TV and movies has sped up, with shorter shot lengths that now average 4 seconds. [16] TV ads have similarly shortened, and digital media make extensive use of jump cuts to reduce the length of natural pauses in speech and pack more content into a short video. The average TikTok video length is about half a minute long,[17] and the recommended Instagram Reel length to "boost engagement" is 7–15 seconds.[18] Podcasts, audiobooks, and other media now offer the option of listening at 1.5x or 2x speed, a practice sometimes referred to as "podfasting." The sole reason for this is to be able to consume more content in a given period of time.

The term *time pressure* involves two elements—an objective component of time shortage and a subjective or emotional component of feeling rushed. [19] Time shortage basically refers to a math problem: We do not have enough hours to carry out all the activities on our to-do list. Feeling rushed, on the other hand, is experiential, the subjective experience that life's pace is hectic, a sense of urgency, constant vigilance about our packed schedules, and a feeling of lack of control.

While it is unclear how actual free time has changed over the last decades, studies on the subjective perceptions of time show that most adults feel frequently or always rushed, although there are significant cultural differences. Nearly half of Americans report feeling that they do not have enough time to do what they want to do, two-thirds say they always or sometimes feel rushed, and half say they almost never feel that they have time on their hands.[20]

Several factors are probably at play. First, we have an almost limitless number of choices available to us for entertainment, news, enrichment activities, leisure, etc., designed to take up our time. Second, there is a constructed cultural value of living a hectic lifestyle, which in some circles can be seen as a status symbol. Third, as noted above, technology has brought an accelerated pace, and we spend so much time on devices, where we can run a complex computation or download an entire movie in mere seconds, that we seem to have developed an expectation about the speed at which the world should move. Fourth, our days are fragmented by interruptions and distractions, which cause a disjointed sense of time.[19] As a result, daily life can become infused with a sense of urgency, impatience, and rush.

While we have normalized rushing in our society, chronic time pressure—and its subjective component of feeling rushed in particular—triggers the stress response. Unsurprisingly, over the long term, unrelenting time pressure has significant negative impacts on our emotional, relational, and cognitive health, and it has been associated with increased physical fatigue, high levels of emotional exhaustion, poor sleep quality, and cognitive lapses.[21]

An ongoing rushed pace is not just detrimental for the person rushing; it also affects relationships. You might remember from Chapter 4 that because our brains "read" other people's emotions automatically, stressful emotional states can be transmitted between individuals. Children of parents who experience frequent time pressure are more likely to display emotional and behavioral problems than children whose parents experience time pressure only occasionally.[22] It is not the objective shortage of time that is associated with the children's mental health, but the parents' stressful subjective experience of rushing: Time pressure is a stressor for the parents, and the parents' stress is a stressor for the children.

The urgency to go faster is not an intrinsic property of our brains. Our brains evolved the capacity to process an incredibly large amount of information very fast, and to do so in the most metabolically and cognitively efficient ways. But our brains do not crave speed; the emphasis on speed is a cultural value. It is constructed and reinforced by societies, and not by all societies. Many cultures value slower personal tempo, deliberate thought, and meandering conversation over fast pace, efficient outcomes, and to-the-point interactions.[23] Rushing is, to a large degree, a collective choice.

Social (Dis)Connection

Studies of people's estimation of hill slopes have shown that we see a hill as steeper when we are wearing a heavy backpack, when we are fatigued... and when we are alone. Research participants who are accompanied by a friend perceive a hill as less steep than those who are alone, and the effect is stronger the better the quality of their relationship.[24] A hill seems easier to climb when we are with a supportive loved one: We literally perceive the world differently when we are together.

Yet, globally, half of adults report some degree of loneliness.[25] Against common assumptions, the rates of loneliness are higher for young than for older adults. In the U.S., where similarly half of adults report feeling lonely, time spent alone has increased over time, and the number of close friend-ships has declined, with a worsening of these trends during the Covid pandemic.[26] Two-thirds of Americans say they need more emotional support than they receive, and more than half say they wish they had someone to turn to for advice or support.[27]

We experience loneliness based not just on the amount of time we spend around others, or the number of people we are around, but the quality of those connections and how satisfied we feel with our social connectedness. While social *isolation* refers to spending large amounts of time alone and having infrequent interactions, we feel *lonely* when we feel disconnected, regardless of whether we are with others or not. A person might feel lonely not because they do not have enough relationships, but because their relationships are not as deep or close as they would like.

Social connection is immensely important for brain health and health in general. Loneliness has been associated with increased risk of cardiovas-cular disease including stroke, metabolic syndrome, disability, poor health behaviors, poor sleep, depression, anxiety, decreased well-being, impaired executive control, suicidal ideation, and increased mortality.[28] Persistent loneliness in midlife has been associated with cognitive decline, increased risk for dementia, and decreased volume in brain regions important for memory and executive functioning.[29], [30] In fact, loneliness is con-sidered so bad for our health that the effect of loneliness on mortality has been estimated to be the equivalent of smoking 15 cigarettes a day, well above that of physical inactivity,[26] and has led to loneliness being thought of as a public health concern.

On a day-to-day level, based on the hill slant studies, loneliness likely makes us see the difficulties in our life as more challenging, perhaps insur-mountable, significantly exacerbating cognitive overload. Moreover, our brains evolved within the uniquely human social context, and were shaped by our interactions with others, by our interdependence. We can think of our close social network as one more neural network. Having a healthy social network is as important to our functioning as having healthy atten-tional, executive, or memory networks. We are not meant to work, parent, or care for ourselves and others in isolation. Our brains are not wired to carry out their mission of keeping us safe and well by themselves.

* * *

We have now reviewed many forces that, together, deplete our internal resources and overload us with excessive external demands. The remaining chapters of this book will go over specific actions that can help us reclaim our cognitive resources and build brain-friendlier days.

Resources

- For more on the topic of technology's impact on our brains and behavior, *Irresistible: The Rise of Addictive Technology and the Business of Keeping Us Hooked*, by Adam Alter, reviews the human and technological factors that fuel behavioral addictions to social media, gaming, work, shopping, exercise, and others.
- Johann Hari's book *Stolen Focus: Why You Can't Pay Attention—and How to Think Deeply Again* explores how social and other digital media is designed to capture our attention by basing their designs on neuroscientific knowledge and disregarding its psychosocial effects. Importantly, the book also focuses on the crucial systemic factors and societal-level solutions that I do not address in this book.
- The website for Tristan Harris and Aza Raskin's Center for Humane Technology (www.humanetech.com) has information and resources regarding technology use. They also host a podcast, Your Undivided Attention, where they discuss topics related to the power of emerging technologies, how they affect our lives, and how best to use them.
- The comprehensive U.S. Surgeon General report on the health effects of loneliness can be found at www.hhs.gov.

References

1. Laricchia, F. (2023, December 6). Daily time spent on mobile phones in the U.S. 2019–2024. *Statista*. www.statista.com/statistics/1045353/mobile-device-daily-usage-time-in-the-us.
2. Nielsen. (2018, July). Time Flies: U.S. Adults now spend nearly half a day interacting with media. www.nielsen.com/insights/2018/time-flies-us-adults-now-spend-nearly-half-a-day-interacting-with-media.
3. Gelles-Watnick, R. (2024, January 31). Americans' use of mobile technology and home broadband. Pew Research Center. www.pewresearch.org/internet/2024/01/31/americans-use-of-mobile-technology-and-home-broadband.
4. Center for Humane Technology. (2022). Attention & Mental Health. www.humanetech.com/attention-mental-health.
5. Dixon, S.J. (2024). Social media—Statistics & Facts. *Statista*. www.statista.com/topics/1164/social-networks/#topicOverview.

6. Rothwell, J. (2023, October 13). Teens spend average of 4.8 hours on social media per day. *Gallup*. https://news.gallup.com/poll/512576/teens-spend-avera ge-hours-social-media-per-day.aspx.

7. Hancock, J., Liu, S., Luo, M., & Mieczkowski, H. (2022). Psychological well-being and social media use: A meta-analysis of associations between social media use and depression, anxiety, loneliness, eudaimonic, hedonic, and social well-being. Available at SSRN: 4053961.

8. McComb, C.A., Vanman, E.J., & Tobin, S.J. (2023). A meta-analysis of the effects of social media exposure to upward comparison targets on self-evaluations and emotions. *Media Psychology*, 26(5), 612–635.

9. Mark, G., (2023). *Attention span: Find focus, fight distraction. A groundbreaking way to restore balance, happiness, and productivity*. Hanover Square Press.

10. International Labour Organization. (2022). *Working time and work-life balance around the world*. International Labour Office.

11. American Psychological Association. (2023). 2023 Work in America survey. www.apa.org/pubs/reports/work-in-america/2023-workplace-health-well-being.

12. Hansen, K., & Saini, A.N. (2020, July 15). A brief history of the modern office. *Harvard Business Review*. https://hbr.org/2020/07/a-brief-history-of-the-modern-office.

13. U.S. Bureau of Labor Statistics. (2023, June 22). Economic news release: American time use survey summary. www.bls.gov/news.release/atus.nr0.htm.

14. Mills, K. (Host). (2021, June 30). Back to the Office? The future of remote and hybrid work, with Tsedal Neeley, Ph.D. [Audio podcast episode]. Speaking of Psychology. American Psychological Association.

15. Tomlison, J. (2007). *The Culture of Speed*. Sage.

16. Mills, K. (Host). (2023, February 8). Why our attention spans are shrinking, with Gloria Mark, Ph.D. [Audio podcast episode]. In *Speaking of Psychology*. American Psychological Association.

17. Ceci, L. (2023, November 14). Average TikTok video length from March 2023 to August 2023, by number of video views. *Statista*. https://www.statista.com/statis tics/1372569/tiktok-video-duration-by-number-of-views/.

18. Sharethis. (2022, November 17). The best Instagram Reels length to boost engagement. https://sharethis.com/social-media/2022/11/best-instagram-reels-leng th-to-boost-engagement/.

19. Szollos, A. (2009). Toward a psychology of chronic time pressure: Conceptual and methodological review. *Time & Society*, 18(2–3), 332–350.

20. Rudd, M. (2019). Feeling short on time: Trends, consequences and possible remedies. *Current Opinion in Psychology*, 26, 5–10.

21. Kleiner, S. (2014). Subjective time pressure: General or domain specific? *Social Science Research*, 47, 108–120.

22. Gunnarsdottir, H., Bjereld, Y., Hensing, G., Petzold, M., & Povlsen, L. (2015). Associations between parents' subjective time pressure and mental health pro- blems among children in the Nordic countries: A population based study. *BMC Public Health*, 15, 353.

23. Fujii, D. (2016). *Conducting a culturally informed neuropsychological evaluation*. American Psychological Association.

24. Schnall, S., Harber, K.D., & Stefanucci, J.K. (2008). Social support and the per- ception of geographical slant. *Journal of Experimental and Social Psychology*, 44 (5), 1246–1255.

25. Gallup & Meta. (2023). *The global state of social connections*. Gallup.

26. Office of the U.S. Surgeon General. (2023). *Our epidemic of loneliness and isolation. The U.S. Surgeon General's advisory on the healing effects of social connection and community.* U.S. Department of Health and Human Services.

27. American Psychological Association. (2023). Stress in America 2023: A nation recovering from collective trauma. www.apa.org/news/press/releases/stress/2023/collective-trauma-recovery.

28. Hawkley, L.C. (2022). Loneliness and health. *Nature Reviews Disease Primers, 8,* 22.

29. Tao, Q., Akhter-Khan, S.C., Ang, T.F.A., DeCarli, C., Alosco, M.L., Mez, J., Killiany, R., Devine, S., Rokach, A., Itchapurapu, I.S., Zhang, X., Lunetta, K.L., Steffens, D.C., Farrer, L.A., Greve, D.N., Au, R., & Qiu, W.Q. (2022). Different loneliness types, cognitive function, and brain structure in midlife: Findings from the Framingham Heart Study. *The Lancet eClinicalMedicine, 53,* 101643.

30. Salinas, J., Beiser, A.S., Samra, J.K., O'Donnell, A., DeCarli, C., Gonzales, M.M., Aparicio, H.J., & Seshadri, S. (2022). Association of loneliness with 10-year dementia risk and early markers for vulnerability for neurocognitive decline. *Neurology, 98,* e1337–e1348.

11 Slowing Down

My first job as a faculty member was with a family medicine residency program, teaching residents about the psychosocial aspects of health. I shadowed the residents in clinic, modeling ways to address mental health concerns and giving them feedback on their interpersonal skills. During my time there, I worked with a resident, Valeria. She was a kind and thoughtful physician, well liked by patients, but the faculty had concerns about her ability to manage the heavy workload and her overall professional resilience.

Late one afternoon during a particularly busy clinic, one of our attendings came to me quite frustrated. They had just seen Valeria, who was already running behind (like everybody was almost every day), standing outside an exam room, looking over her next patient's chart, *eating a peach*. My colleague was appalled. It was almost a half-hour after the patient's scheduled appointment, and Valeria was just standing there, eating. Another colleague, hearing this, shook their head in disappointment.

I cringed at the image. Phones ringing, staff running around, babies crying, and Valeria eating her peach. When I spoke to her about it, she calmly said, "I had been rushing all day, I hadn't eaten, I felt frazzled. So I took *one* minute to eat a snack and breathe." Her transgression—the behavior perceived as unprofessional and unbecoming of a physician—was to *slow down*. Worse yet, she had slowed down to meet her own needs. I was expected to talk to her about promptness, duty, and self-sacrifice, but she had actually done exactly what I routinely advised my patients to do when overwhelmed. Was her behavior a problem, or was she the one acting sanely in an unreasonable environment, refusing to play along with our shared delusion that a frantic pace was the only way?

A World at 1.5x Speed

In our fast-paced world, some degree of time pressure is probably inevitable, and when we are not depleted, it can actually have a positive, motivating effect under certain circumstances. Especially when the task is challenging but achievable, a time crunch can help us "rally" and can even feel deeply

DOI: 10.4324/9781003409311-15

satisfying. Other times, it can make us feel paralyzed and helpless. One of the factors that determines whether time pressure is motivating or hindering is how long we experience it.

As we mentioned in Chapter 10, time pressure triggers our stress response. [1] Like other stressors, in short bursts time pressure can energize us and help us mobilize and focus our mental resources, but, when chronic, its effects can instead be detrimental and demotivating. At work, for example, elevated time pressure lasting a day or a week increases employee engagement, but time pressure extending over weeks decreases engagement.[2]

We also mentioned that a constantly rushed pace is not just bad for ourselves but hurts our relationships. Think about what it feels like to interact with someone who is rushing: a physician that seems eager to get out of the room and asks, "Any questions?" with their hand already on the doorknob; a teacher who continues to walk fast as you ask a question about your child; the coworker showing you around on your first day who goes over everything quickly and starts answering before you have finished asking your questions. Interacting with someone in a rush can make us feel unseen, dismissed, and disconnected.

When We Rush Our Brains

In a culture that values speed and efficiency, we can become frustrated with what we perceive as our inability to go fast enough. The truth is that there are large individual differences in cognitive processing speed: Among cognitively healthy individuals, "normal" processing speeds—meaning how fast most healthy individuals can process information, excluding those who are exceptionally slow or exceptionally fast—encompass a very wide range.

For example, for a typical 50-year-old, "normal" performance on tests of processing speed could mean obtaining a score anywhere between 30 and 70 percent.[3] This means that, among healthy adults, a certain pace of mental activity that is easily manageable for one person can be unmanageable for another, not because there is anything wrong with them, simply because their perfectly healthy brain does not process information that fast. And remember that processing speed is sensitive to the effects of sleep deprivation, substance use, medication side effects, depression, and other factors, and it is one of the cognitive functions that declines pronouncedly with age.

Moreover, chronic rushing is not just unsustainable and detrimental for us and those around us, *it does not work*. Even under the best of circumstances, there is a speed–accuracy trade-off. It takes our brains time to gather and process information, especially for complex activities like making a decision. Our brains have networks that balance the competing demands of speed and accuracy. Prioritizing speed causes more errors even on simple tasks and increases the baseline activity of neurons in those control networks: We make our brains work harder by rushing.[4]

Think about how many times you have had to go back and correct mistakes you made when rushing, effectively erasing any time you had saved. How many times have you fired an email in a rush, only to realize after you hit "Send" that you misread something or left something important out, so you have to send a second email? How many times have you hurried out of the house in the morning, only to turn around when you realized you left your child's lunchbox or your work documents on the kitchen counter? I recently spent hours fixing a problem I caused by filling out a form in a hurry and entering the wrong bank account number.

We also rush when there is no need to. Our perception of time is fallible, and it can be distorted by our mental state. When we are in a negative emotional state, when we are dealing with multiple complex activities, and when we are highly invested in an outcome, we perceive time as slowing down or expanding, a phenomenon known as time dilation.[5] This can lead us to overestimate how long things will take and feel we have to rush even if we actually have enough time.

On top of that, we tend to overestimate how much time we will save by going faster. The so-called time-saving bias has been related to, for example, people's willingness to speed, even though speeding usually saves us a negligible amount of time:[6] The average driver in an urban setting saves 26 seconds a day and two minutes a week by speeding, at an increased risk of injury and death.[7]

Finally, remember that processing speed is intimately related to other cognitive functions like attention, memory, and problem-solving. Because of this, processing speed deficits can have wide impact on activities of daily life. If we are depleted and processing information more slowly than usual, while the world around us continues at its swift pace, we are more likely to miss things, make mistakes, and misremember information at a later time. This might lead us to think there is something wrong with our memory, when what we have is a processing speed problem due to incompatibility between our slowed-down speed and the speed of the world. The same glitches will occur even if there is nothing slowing our processing speed down, if we rush through our day so that information and activities are coming at our brain faster than it can process them.

What You Can Try Today

1. Acknowledge your natural pace. The first step in any behavioral change is increased awareness. We all have a natural mental pace: Think about how fast you talk, read, and type when you are not rushing. On top of this baseline pace, there might be things going on today slowing you down:

- How was your sleep last night? Do you feel rested?
- How is your stress level today? Do you feel calm, tense, restless, or scattered?

- Is your mind busy with thoughts about things you are preoccupied with, or is it quiet up there?
- How is your mood today? Do you feel sad, down, or anxious?
- Are you in pain or other physical discomfort?
- Have you taken any medications that make you feel sluggish?
- Did you use (perhaps over-use) any substances last night?

If you are depleted, your processing speed is slowed down, and you find yourself overwhelmed or confused because it feels that the world around you is going too fast, there is really only one way to manage that. First, acknowledge your brain's current speed as a natural consequence of the factors draining its resources. Accept that this means you have to go at that slower pace, which likely means you will accomplish less than when you are not depleted. Second, do not try to rush yourself: We cannot make our brain go faster than its limit at any moment, any more than we can make our legs go faster than their limit, at least not without tripping. Instead of fighting your brain over its tempo, accept it and work with it. Finally, try to slow your day down to match your brain's speed by using the strategies below.

If you are not particularly depleted or slowed down, but you still find yourself overwhelmed and making mistakes because you are rushing through your day due to the sheer volume of your to-do list, these same strategies can help you slow down to avoid errors and glitches.

2. Notice when you are rushing. Throughout the day, remind yourself to turn your attention to whether you are experiencing that subjective sense of time pressure. Notice how rushing comes with tension. Notice how your body feels when you're rushing and when you slow down. Notice when you are rushing others, and potentially spreading your tension around. Remember that our perception of time can be biased, and ask yourself, "Am I really in a hurry right now?"

3. Check. Then double-check. When depleted and when rushing, we make mistakes we do not typically make. Take time to proofread that memo you just finished, double-check the amount on that payment before submitting, and confirm that you got everything on your grocery list before heading home. Remind yourself that rushing rarely saves a meaningful amount of time, and that in fact you might be saving time by slowing down and intentionally focusing on what you are doing.

4. Prepare and plan. When our processing speed is slow, we can struggle in situations when we have to think, answer questions, or make decisions quickly. Our doctor asks if we have any other questions and we cannot think of any (but remember them later), or we have difficulty quickly scanning the menu and deciding what to order in a busy restaurant. So prepare: Take notes before your doctor's appointment of important things you want to cover; review the restaurant's menu and decide what you want to order in advance.

5. Give slowness a try. If you are late leaving the house and you feel yourself rushing, notice how it feels to slow down as you put your shoes on and gather your things. Get on the right lane while driving. Slow your pace walking from your car to the building. Test for yourself how much difference it makes, in terms of time, to slow down. Notice the cues to rush, all around you—the car tailgating you, then brusquely passing you; the person doing the same on a narrow sidewalk.

Early one morning, leaving a class at the nearby YMCA, I realized I had scheduled two back-to-back remote meetings with students, the first one starting in seven minutes. I could have rushed home berating myself for my absent-mindedness, hopped on the videocall in my workout clothes, and not eaten breakfast for another two hours. Instead, I chose to run a little experiment: I messaged my student to let her know I would be a few minutes late. Once home, I timed myself as I washed my face, changed my clothes, and grabbed a snack. It took me *four* minutes to do this; that is how much time rushing would have saved me. Instead of being three minutes late, I was seven minutes late, but I did not feel rushed, I was not distracted by hunger or self-critical thoughts, and I was fully present.

6. *Choose wisely.* Slow down selectively. The goal is not to go through the day in slow motion. Remember that it is the *subjective* feeling of time pressure that is counterproductive. There are many things we can do fast, especially things that are overlearned or routine. Anything that you can do fast and well without feeling pressured or stressed is not a problem (maybe "podfasting" works for you). But by reducing the baseline speed of your days, you will be able to better handle those inevitable times when you are rushed. Remember, our brains can handle short bursts of time pressure, as long as we are able to return to a manageable baseline.

7. *Say it out loud.* Normalize slowing down. When others are rushing, ask them to slow down or repeat themselves. If a meeting is moving too fast, say, "Can we slow down a bit? This is important." Say to the pharmacist rushing through your prescription information, "I need you to repeat that, please. I want to make sure I don't miss anything." And if it is you rushing, call yourself out: If you catch yourself telling your children about the plans for a busy day too fast, say, "Wait, I'm throwing a lot at you really fast. Let me start over and go slower." By doing this, you are conveying that what is being said is worth taking in carefully, and you are modeling how to do that by slowing down.

References

1. Rojo López, A.M., Cifuentes Férez, P., & Espín Lopez, K. (2021). The influence of time pressure on translation trainees' performance: Testing the relationship between self-esteem, salivary cortisol, and subjective stress response. *PLoS ONE*, 16(9), e0257727.
2. Baethge, A., Vahle-Hinz, T., Schulte-Braucks, J., & van Dick, R. (2018). A matter of time? Challenging and hindering effects of time pressure on work engagement. *Work & Stress*, 32(3), 228–247.

3. Wechsler, D. (2008). *WAIS-IV Administration and scoring manual*. NCS Pearson.
4. Bogacz, R., Wagenmakers, E.-J., Forstmann, B., & Nieuwenhuis, S. (2009). The neural basis of the speed-accuracy tradeoff. *Cell*, 33(1), 10–16.
5. Sucala, M., & David, D. (2012). Slowing down the clock: A review of experimental studies investigating psychological time dilation. *The Journal of General Psychology*, 139(4), 230–243.
6. Eriksson, G., Svenson, O., & Eriksson, L. (2013). The time-saving bias: Judgements, cognition and perception. *Judgment and Decision Making*, 8(4), 492–497.
7. Ellison, A.B., & Greaves, S.P. (2015). Speeding in urban environments: Are the time savings worth the risk? *Accident Analysis and Prevention*, 85, 239–247.

12 Hitting "Pause"

Sure, you left the house with two different shoes on because you were rushing. You said, "Bye, love you!" to your child's math tutor because you were thinking about your partner's upcoming work trip. You might have forgotten that you were supposed to buy the present for Boss's Day with the money your coworkers collected. But chances are you have not taken out the wrong kidney during surgery lately.

In 2004, the Joint Commission, an organization that accredits and certifies healthcare organizations, instituted the *Universal Protocol for Preventing Wrong Site, Wrong Procedure, Wrong Person Surgery*.[1] The protocol includes a mandatory, standardized time-out: Immediately before making a surgical incision or starting an invasive procedure, the surgical team goes through a checklist to ensure there is agreement about the patient's identity, the site of the procedure, and the specific procedure to be performed. The timeout is also an opportunity for any team member to raise questions or concerns, and the procedure is not started until those have been addressed. This protocol was instituted based on two decades of data on wrong-site surgeries, which, though extremely rare (at the time, one out of every 113,000 surgeries), could have devastating consequences, like the amputation of the wrong limb.[2]

Even the best and brightest among us need to pause.

When We "Brain" Non-Stop

Like rushing, not taking breaks does not work. Our ability to maintain focused attention decreases the longer we are engaged in a task, and our likelihood of making mistakes increases.[3] We start getting distracted by other activities (like checking social media or email) or unrelated thoughts (like worrying or planning). Our top-down cognitive control, meaning our ability to direct and maintain our attention and other cognitive resources to the task at hand, decreases.[4] Remember that our brain is wired to react to novelty: We have a salience network because it is important to our survival to detect changes in the environment. This vital feature is just doing its job when it engages in task-unrelated thoughts or stimuli, and we can't really turn it completely off.

DOI: 10.4324/9781003409311-16

Not taking a pause also affects our memory. One reason memories fade is because we do not rehearse them—we do not use them, so we lose them.[5] When we are going from one task to another without processing information further, we are less likely to remember them at a later time. The *retrieval practice effect* refers to the fact that testing people on information they are trying to learn, so they have to actively retrieve the information or practice it, solidifies their learning. This is why, when preparing for a test, testing yourself (e.g., by answering questions about the material) is more helpful than simply looking the information over multiple times. Pausing during the day to rehearse information helps us remember what we did and heard.

Moreover, we mentioned in Chapter 7 that sleep helps with the consolidation of memories—their "archiving" into long-term memory. Consolidation does not just happen during sleep, and with rehearsal, but it also occurs when the hippocampus is inactive, free from interference, like during a true pause.[6]

What Is a Good Break?

Much of the literature on breaks comes from studies done in work settings. Regular breaks at work, like going for a quick walk, help maintain concentration.[7] A "good" break is one that serves a *recovery* function, and this happens when it allows us to both rest and recharge.[8], [9] In other words, during a good break we:

a   remove task demands, meaning we disengage from doing tasks and thinking about tasks; and
b   replenish our psychological resources, meaning we relax physically and mentally, and ideally induce some positive mood.

Other things that help are when we get to decide when to take a break, when we spend our break in natural environments, and when we take multiple short breaks of even five or ten minutes (although longer breaks are needed when we are engaged in more cognitively demanding tasks).[9], [10], [11] Interestingly, spending our break solely taking care of basic needs, like eating, does not bring the same benefits.

A break is *not* an invitation to shift to another task: Stopping a task to return a phone call or run a quick errand is not a restorative pause. In a true pause, we are either cognitively disengaged (when doing stretches or listening to a calming song) or engaged in cognitively effortless, time-limited activities (a quick call with a friend or reading a brief, uplifting poem).

What You Can Try Today

1. Figure out what you need. As you are going through your day and start feeling tense, fatigued, overwhelmed, or notice yourself glitching, stop and check: Do you feel tense and need to relax? Do you feel fatigued and need to increase your alertness? Do you feel scattered and need to re-focus? Will you be spending the rest of your day doing monotonous, boring things, or shifting among many stressful activities? This will guide how to best spend your breaks.

2. Decide when to take breaks. There are two ways to take breaks: After a certain period of time (e.g., a ten-minute break every hour), or at natural break points (e.g., every time you finish one of the sections of the long report you are writing). What works best will depend on the state of your cognitive "battery" that day and what you are doing. Do what will require less effort and mental resources.

If you decide to take breaks based on time, set a timer so you do not have to keep checking the clock. You might have heard of the Pomodoro technique, which usually consists of taking a five-minute break every 25 minutes (there are timers and apps for this). But there are individual differences, and if you are depleted physically and/or cognitively, you might need even more frequent breaks. Some days you might be able to focus for an hour at a time, other days only for 15 or 30 minutes before needing a few minutes "off." There are no rules; do what helps.

Taking breaks at natural points in what you are doing, or in between tasks, is particularly helpful because it reduces shifting or helps mark the shift between one task and the next. Make your next break point explicit and decide in advance what you will do during your break ("I'm going to finish filing these forms, then I'll walk outside for five minutes before I start catching up with emails").

3. Remember your body. Especially if our days tend to be sedentary, moving our body (by stretching or walking around) is one of the best things we can do during our pauses.

A special kind of exercise break is worth mentioning: Engaging in aerobic exercise of at least moderate intensity for 20 or 30 minutes improves memory performance on a subsequent test.[12] Older adults, in particular, benefit when they exercise before hearing information to be remembered or after hearing the information. So taking an exercise break before or after an activity when you receive important information (like your first diabetes education class or a meeting with your accountant) is a great idea.

Exercise breaks also reduce physiological stress markers and increase energy levels, while relaxation is better at helping us detach from tasks.[7] So if you need to increase alertness, exercise; if you need to reduce mental overwhelm, do a relaxation exercise (see Chapter 19).

4. Seek something positive. Positive emotion can help replenish our cognitive energy. Watch a short heart-warming or funny clip, look at pictures of your family, or browse through beautiful nature photographs. This can be tricky to do online, because your attention can get hijacked or you might be exposed to cognitively expensive content (like negative headlines), so choose your source carefully and set a timer if needed.

5. Seek the light. We will see in Chapter 20 that natural light is one of the most effective ways to increase alertness. Being outside has positive effects on psychological health, especially when we are engaged in ongoing tasks that take most of our day on most days, like working long hours, parenting, or caregiving. Being outside can be particularly helpful when it combines physical activity, bright light, and positive emotional experiences (like a beautiful view of the mountains, birds collecting sticks for their nests, or sunlight over a pond). For example, walking in a park is more restorative and decreases fatigue more than walking in an urban area.[8] But you can also stand by a window to do breathing exercises or stretches.

6. Do nothing. Sometimes, it is best to rest and let our minds wander. Do not look at your phone, read, or chat with your favorite coworker. Do not give your brain anything to process. Remember the default mode network: Stepping away from cognitive tasks results in self-referential thoughts about your past and future, but also the development of connections between contents that were previously unrelated in your mind, and that can help with memory and creativity.

7. Take a nap. We will see in Chapter 20 that taking a short "cat nap" is an effective way to increase alertness. Longer naps, during which we go into deep sleep, can have positive effects in executive functioning, memory consolidation, learning, and emotional processing, but they cause *sleep inertia*, meaning we feel drowsy when we wake up.[6]

If you are sleepy and need to become more alert, take a short, 10–15-minute nap to avoid going into deep sleep and waking up drowsy. But if you just came back from a meeting with an insurance broker or your child's teacher where you received a lot of new information (or if you are going to such a meeting later in the day), take a longer nap as long as you have time to recover from sleep inertia when you wake up.

8. Connect with a supportive one. Time with loved ones is one of the things that we often neglect when we are depleted or overwhelmed. We often mistakenly think we do not have enough time to meaningfully connect with one another, but consider this: Physicians often interrupt patients because they are pressed for time, they feel they have to keep the appointment focused, and they fear patients will speak for too long if not re-directed. But a classic study found that, if uninterrupted, patients only spoke for an average of one and a half minutes; 80 percent were "done" talking after two minutes.[13] It does not take long to feel a connection, to feel heard and help others feel heard. Text a friend, "I'm going to take a quick ten-minute break. Do you want to touch base for a bit?" Or "I'm going on a five-minute walk. Can I call you to vent about something?"

9. Implement an "executive pause." This strategy is not for the kind of restorative pauses we have been discussing, but for what I think of as an "executive pause," critical when we are depleted. This kind of intentional self-monitoring is so helpful that it is recommended for neurological patients, like those recovering from a brain injury, who are experiencing executive problems.[14] As mentioned above, taking a break in between tasks to re-orient yourself and mark the shift from one task to the next can significantly reduce glitching and errors. Pause to signal to your brain, explicitly and calmly, what you are leaving behind and what you are starting next.

Also use pauses to organize your mental archives: After you have received a large amount of important information, pause to allow your brain to encode it ("record" it) better. Remember the retrieval practice effect. After a surgical appointment or an orientation session about the upcoming switch in software systems at work, pause to rehearse important information so you enhance your memory of it: What did the surgeon say were the possible risks of the surgery? When did the manager say the software is launching? Use mnemonic strategies as you rehearse, like word associations ("The hospital bill is due May 1st. In May we pay!") and imagery (to remember your new surgeon is Dr. Wolf, you can imagine the Big Bad Wolf dressed in scrubs and holding the scalpel in the operating room).

10. Talk about it. Normalize pausing instead of jumping from one thing to the next. At the end of a fast-paced meeting with your boss, say, "Let me repeat back to you what you said, I want to make sure I'm not missing anything." You can even pause in the middle of a conversation, when you are feeling overwhelmed, and say, "Give me one minute to process this." If you are on an upsetting phone call and start having difficult regulating your emotions, say, "I need to take a quick break, I'll call you right back." If you are picking up your children from one activity and driving to another, say, "Let's just sit here in the car and listen to a song quietly before we leave."

References

1. The Joint Commission. (undated). The Universal Protocol. www.jointcommission.org/standards/universal-protocol.
2. Kwaan, M.R., Studdert, D.M., Zinner, M.J., & Gawander, A.A. (2006). Incidence, patterns, and prevention of wrong-site surgery. *Archives of Surgery*, 141, 353–358.
3. Helton, W.S., & Russell, P.N. (2012). Brief mental breaks and content-free cues may not keep you focused. *Experimental Brain Research*, 219, 37–46.
4. Ariga, A., & Lleras, A. (2011). Brief and rare mental "breaks" keep you focused: Deactivation and reactivation of task goals preempt vigilance decrements. *Cognition*, 118(3), 439–443.
5. Schacter, D.L. (2022). The seven sins of memory: An update. *Memory*, 30(1), 37–42.
6. Mantua, J., & Spencer, R.M.C. (2017). Exploring the nap paradox: Are mid-day sleep bouts a friend or foe? *Sleep Medicine*, 37, 88–97.
7. Diaz-Silveira, C., Santed-Germán, M.-A., Burgos-Julián, F.A., Ruiz-Íñiguez, R., & Alcover, C.-M. (2023). Differential efficacy of physical exercise and mindfulness during lunch breaks as internal work recovery strategies: A daily study. *European Journal of Work and Organizational Psychology*, 32(4), 549–561.
8. De Bloom, J., Sianoja, M., Korpela, K., Tuomisto, M., Lilja, A., Geurts, S., & Kinnunen, U. (2017). Effects of park walks and relaxation exercises during lunch

breaks on recovery from job stress: Two randomized controlled trials. *Journal of Environmental Psychology*, 51, 14–30.

9. Lyubykh, Z., Gulseren, D., Premji, Z., Wingate, T.G., Deng, C., Bélanger, L.J., & Turner, N. (2022). Role of work breaks in well-bring and performance: A systematic review and future research agenda. *Journal of Occupational Health Psychology*, 27(5), 470–487.

10. Albulescu, P., Macsinga, I., Rusu, A., Sulea, C., Bodnaru, A., & Tulbure, B.T. (2022). "Give me a break!" A systematic review and meta-analysis on the efficacy of micro-breaks for increasing well-being and performance. *PLoS One*, 18(8), Article e0272460.

11. Kim, S., Park, Y., & Headrick, L. (2018). Daily micro-breaks and job performance: General work engagement as a cross-level moderator. *Journal of Applied Psychology*, 103(7), 772–786.

12. Mills, K. (Host). (2022, April 20). How exercise benefits the brain, with Jenny Etnier, Ph.D. [Audio podcast episode]. Speaking of Psychology. American Psychological Association.

13. Langewitz, W., Denz, M., Keller, A., Kiss, A., Rüttimann, S., & Wössmer, B. (2002). Spontaneous talking time at start of consultation in outpatient clinic: Cohort study. *BMJ*, 325, 682–683.

14. Jeffay, E., Ponsford, J., Harnett, A., Janzen, S., Patsakos, E., Douglas, J., Kennedy, M., Kua, A., Teasell, R., Welch-West, P., Bayley, M., & Green, R. (2023). INCOG 2.0 guidelines for cognitive rehabilitation following traumatic brain injury, part III: Executive functions. *Journal of Head Trauma and Rehabilitation*, 38(1), 52–64.

13 Taming Our Attention

Attention can feel like a feral animal, indomitable and unpredictable. Like a puppy failing obedience school, it can jump all over the place chasing one thing after the next, or become fixated on a single, incomprehensible, irrelevant thought. What is a reasonable expectation for attentional control, in particular when your cognitive resources are depleted?

When Attention Is Untamed

As we reviewed in Chapter 2, attention actually involves several processes: We need to be able to focus attention, maintain attention, divide attention, and selectively focus our attention. More effortful and complex, executive aspects of attention include top-down attentional control, inhibition of distractions, and the flexible shifting of attention.[1] There are a few aspects of attention worth reviewing because they are relevant to the strategies that help tame our emotion.

First, there is often tension between bottom-up, stimulus-driven attentional processes and the top-down, goal-directed, executive control of attention.[2] The salience network in our brains is designed to detect and direct attention to information in the environment that could be relevant to our safety and well-being, while executive systems filter out most information coming through our senses and allocate our attentional resources to the task we are prioritizing at that time. For example, while sitting at a lecture, our salience network detects our phone buzzing in our pocket, the whispers of the people sitting behind us, and our craving for a sweet snack, while top-down attentional control inhibits those sensory stimuli and maintains our attention on the voice of the speaker and the information on the screen.

Second, as we saw in Part II, attention is prone to disruption by mood or anxiety problems, health problems including pain, substances and medications, and sleep loss. When we are depleted, running those other "apps," there is a perfect storm of distractibility: Our top-down executive control is weakened, while at the same time stress enhances our sensory scanning because our brain is programmed to pay attention to the environment to

DOI: 10.4324/9781003409311-17

detect threats. We are more distractible *and* we have fewer resources to control that distractibility.

Finally, attention is the gateway to consciousness and cognition: What is not consciously attended to cannot be consciously used and will not be consciously remembered (I emphasize *consciously* because we do process and "remember" information unconsciously, as we reviewed when discussing trauma). Many memory failures are not due to a memory problem but an attention problem: We are absent-minded, preoccupied, ruminating, or lethargic, so we are distracted when our brain is encoding (learning) or recalling (retrieving) information.[3] If we are distracted when we set our keys down, we will later not know where they are, not because we forgot where we left them, but because we never noticed where we left them. Similarly, if we are distracted when talking to someone, we might call them the wrong name, not because we forgot their name, but because we were distracted when retrieving it.

"I'm So ADD"

Your glitching might lead you to wonder if you have attention-deficit/ hyperactivity disorder (ADHD). ADHD is condition characterized by a pattern of

a inattention—for example, difficulty maintaining attention, poor attention to detail, and organizational problems; and/or
b hyperactivity and impulsivity—for example, restlessness, blurting out answers before a question is completed, and frequently interrupting others.

The term "ADD" (attention deficit disorder) is sometimes used to refer to people who experience problems with attention but not hyperactivity.

ADHD is a neurodevelopmental disorder, meaning the symptoms begin in childhood. While the term is sometimes used flippantly to refer to a tendency to be easily distractible, the diagnosis is actually complex and is associated with negative consequences in terms of health and psychosocial outcomes, including an increased risk of substance use, injuries, employment instability, and relationship problems, among others.[4] It is a disorder needing careful diagnosis and treatment. It is not just distractibility, and it is not an appropriate diagnosis for people whose attention lapses are due to cognitive depletion or overwhelm.

Most children with ADHD will continue to show some symptoms into adulthood, even if the condition improves.[5] Moreover, ADHD can go undetected in childhood and only become obvious when life demands increase in adolescence or young adulthood. It is not that the condition started in adulthood; it just started causing problems in adulthood. Adults with ADHD often struggle with executive functions including attentional control but also organization, planning, goal monitoring (awareness of whether they are still on track to accomplish what they want to accomplish), initiation (getting started on tasks), sustained attention (so they can get side-tracked and start many different things without finishing any), attentional shifting, and others.

ADHD is common, and many of us are walking around with unmanaged or undiagnosed ADHD. But a depleted brain without ADHD can struggle with many of the things adults with ADHD do, which can translate into daily problems with household management (like organizing our finances), work (like meeting deadlines), and even relationships (like difficulty focusing when our partner is sharing something important to them). But there are many strategies that are proven to help those with ADHD that can also help those of us experiencing attention lapses due to depletion of our brain resources and/or excessive cognitive demands.

What You Can Try Today

1. Pay attention to your attention. We all have different attention spans, both in term of capacity (how much we can focus on) and in terms of time (for how long we can focus at a time), and both of these are reduced when we are depleted. Some days, we can devour a long article without our minds wandering; some days, we have to read the same sentence three times. Accept your attention span today, and use it to plan how you will tackle what you need to do.

If, as you start monitoring your attention more closely, you actually have concerns about whether you might have undiagnosed ADHD, the website for Children and Adults with Attention Deficit/Hyperactivity Disorder (www.chadd.org) has comprehensive, up-to-date, and reliable information about attention deficit disorder. You can review this information before addressing your concerns with your primary care physician.

2. Set attention cues. Experiments have shown that attention can be enhanced by periodic "alerting" cues that remind you to concentrate on the task.[4] If you are working on something that will take a while, and your attention keeps wandering, you can set attention cues to bring you back to the task instead of relying on your brain to spontaneously and repeatedly re-focus. For example, you can set a gentle chime sound to go off every 15 minutes (there are apps for this). Choose a particular sound that you would find helpful and not annoying or startling; a single note will do. Every time you hear the sound, it can be a cue to bring your attention to what you are supposed to be doing.

You can also try pairing the sound with a deep breath, and treat it as a micro-reset: A brief deactivation and reactivation of your focus that lasts just a couple of seconds can actually help you maintain alertness and concentration. Think about it as a slow, long blink to refresh our eyes after you have been staring at a screen for too long.

3. Check your environment. Remember that when we are depleted, we are more distractible. Any "attention protection program" begins with our immediate environment. Clear your cluttered desk before you start working; clear the kitchen counter before you start cooking. Find the quietest room in the house and shut the door. Silence your phone. Wear noise-cancelation headphones or play a concentration-friendly melody that will drown environmental noise out. If you are trying to read an article online, use "reader mode" so you are not distracted by ads (if reader mode is available, you will see the icon of a page on the address bar of your browser).

Also remember that when we are anxious, stressed, or depressed, we tend to focus, and get stuck on, negative stimuli, which can then divert more cognitive resources from the task. Scan your environment for negative triggers, like a pile of unopened mail, and remove them from your sight for now. Surround yourself with cues for positivity, like a family picture, the trinket your favorite uncle gave you, or an award you won.

4. Reclaim your top-down control. Remember that our brain has powerful systems to direct our attention based on our goals and what is important to us at any time. Those systems might be depleted, but we still have that potential. What often undermines our attention is that our brain is pursuing a different goal than we are: Our goal is to help our child finish the science project that is due tomorrow, but our brain is trying to keep us safe by sounding the alarm about everything that could go wrong with our upcoming medical procedure. Think about it as your brain being on a different "setting" than the one you want it to be: You are in work setting, family time setting, or caregiving setting, and your brain is in survival or avoidance setting.

To recruit your attentional control systems to serve your goal, be very intentional and explicit about what is important for you at any given time and what you are trying to accomplish right now. When you find yourself getting repeatedly distracted, check the setting of your attention control system: What goal is it serving? Is it distracting you to avoid an anxiety-provoking task, like looking at your bank account? Is it distracting you to protect you, by making you plan for every little thing that could go wrong during your upcoming trip? Next, try to claim your top-down control back by setting an intention: Remind your brain, out loud if needed, "The most important goal for me right now is to have this science project ready by 7 p.m. so we can have a relaxed dinner." When you catch your attention wandering again, take a breath and set your intention again: "My goal right now is to get this project done."

A great resource with many tools for adults struggling with top-down control of attention and other executive functions is the book *Smart but Scattered Guide to Success*, by Peg Dawson and Richard Guare, authors of a series of books about strategies to improve functioning in ADHD.

5. Practice mindfulness. Mindfulness involves multiple skills and practices, but one of its core elements involves intentional, nonjudgmental attention. Feeling scattered can be extremely frustrating: When you have things to do and your mind keeps jumping from one thing to the next, getting nothing done, borrow this basic mindfulness principle: Throughout the day, practice bringing your attention back, over and over, to the one thing you need to be focusing on *now*. Imagine a loving mother bringing her baby back, over and over, when they crawl a bit too far. Just bring it back, kindly and without scolding, to where it needs to be right now.

The key is bringing your attention back, over and over, *without judgment*. You might catch yourself having thoughts about your trouble focusing—maybe unkind thoughts like "What is wrong with me?" and "I'm such a mess," or helpless thoughts like "I can't do this" and "This is too hard." Every time this happens, take another deep, slow breath, and bring your attention back to what you are trying to accomplish. Remind yourself that your brain is depleted and overwhelmed, and that its jumping around is its attempt at protecting you.

We will review more mindfulness-based strategies and resources in Chapter 19, but if you are unfamiliar with mindfulness practices, one practical and accessible introduction is Shauna Shapiro's Intention, Attention, and Attitude (IAA) model of mindfulness. You can find videos of Dr. Shapiro teaching this model online; a brief introductory one can be found on the website of *Greater Good* magazine (https://greatergood.berkeley.edu).

6. Have a "low capacity" plan. In some situations, you will simply not be able to focus for long. If you are very anxious about an 11 a.m. meeting, plan on doing things with few cognitive demands until then, or things that you can accomplish even if you keep getting distracted. You can tackle a series of discrete, brief activities that take only a couple of minutes, like answering routine emails or making brief phone calls to renew a prescription or schedule home maintenance. Similarly, when the children are playing loud videogames and yelling, it will be easier to clean the pantry than put together your financial paperwork for taxes. When you are extremely depleted or in a high distraction environment, only try to do things that are automatic, habitual, and cognitively cheap, requiring little executive control.

7. Talk about it. Normalize attention lapses, protecting your attention, and decreasing the external demands on it. I have gotten into the habit of saying to people, when my mind wanders during a conversation, "Wait—Say that again. I got distracted just now." Model how to protect attention by instituting "quiet time" at home with your partner and children: depending on your children's ages, this can involve 15-minute, 30-minute, or even one-hour-long periods when you all focus on getting things done quietly and without interruption.

References

1. Ponsford, J., Velikonja, D., Janzen, S., Harnett, A., McIntyre, A., Wiseman-Hakes, C., Togher, L., Teasell, R., Kua, A., Patsakos, E., Welch-West, P., & Bayley, M.T. (2023). INCOG 2.0 guidelines for cognitive rehabilitation following traumatic

brain injury, part II: Attention and information processing speed. *Journal of Head Trauma and Rehabilitation*, 38(1), 38–51.
2. Tsotsos, J.K. (2019). Attention: The messy reality. *Yale Journal of Biology and Medicine*, 92, 127–137.
3. Schacter, D.L. (2022). The seven sins of memory: An update. *Memory*, 30(1), 37–42.
4. American Psychiatric Association. (2022). Neurodevelopmental disorders. In *Diagnostic and statistical manual of mental disorders* (5th ed., Text Revision). American Psychiatric Association Publishing.
5. Sibley, M.H. (2021). Empirically-informed guidelines for first-time adult ADHD diagnosis. *Journal of Clinical and Experimental Neuropsychology*, 43(4), 340–351.

14 Outsourcing Mental Tasks

You stand in the middle of your bedroom. "What did I come in here for?"

You pause in the middle of a conversation. "What was I going to say?"

You play a voicemail from your dentist's office. "We're calling about your missed appointment…"

You find the pills you could have sworn you took this morning sitting on the kitchen counter later in the day.

Some of the most common types of "glitches" in daily life involve failures in our working memory and our prospective memory: Thoughts fade from our minds and we "forget to remember" to do things. Luckily, these are also two cognitive functions that are relatively easy to outsource.

Our Mental Notepad

Remember from Chapter 1 that *working memory* refers to our ability to briefly keep information in mind for a short period of time—like remembering why we walked into a room—and manipulating information in our minds—like estimating the tax on a purchase without using a calculator.[1] It is like a mental notepad, where we keep information briefly while we get something done.[2] When we are depleted and experiencing cognitive overload, the capacity of the notepad is significantly reduced: We read the instructions for a frozen meal and throw the box out, only to have to fish it back out to re-read the instructions (maybe more than once). Information fades from our working memory.

Factors that reduce our working memory capacity include many of the depleting conditions we reviewed in Part II, including psychological conditions like PTSD, obsessive-compulsive disorder, ADHD, anxiety disorders, and depression; stress and other negative emotional states; sleep deprivation; and substance use, including the use of alcohol and cannabis.[3] Some of these factors reduce our working memory capacity through neurochemical effects, but another important mechanism is the interference of ruminative thoughts—those repetitive, intrusive, seemingly uncontrollable, preoccupying thoughts. When our working memory is holding thoughts like "I can't do this," "I really need a drink," and "Nobody cares about me," this

DOI: 10.4324/9781003409311-18

leaves less room for information related to our daily activities. Even dieting has been found to decrease working memory capacity because of repetitive thoughts about food and our bodies, and the costly self-monitoring and self-regulation needed to stick to a regimen and suppress cravings, which decrease the cognitive resources available for other tasks. (Notice that this would be true whether a person is dieting to try to lose weight or to adhere to a medical regimen.)

Factors that increase working memory capacity include mindfulness and possibly physical exercise (likely through the reduction of ruminative thoughts), positive emotional states, increased motivation (like when we know we will receive a reward for completing a task), and when the information in working memory is meaningful to us.[3], [4]

When We Forget to Remember

Prospective memory refers to the ability to remember to execute an intended action in the future.[5] We decide now that we have to remember to do something in the future—we need to "remember to remember." First, we set an intention ("I need to pick up my prescription on my way home from work"). Then, while engaged in other tasks (driving, stopping by the grocery store, getting gas), we have to recall that intention and execute it.

Factors that lead to prospective memory failures include multitasking, disruptions to our routines, and interruptions.[6] What does this look like?

- Today is the one day of the week we stay at work late and our spouse picks up the kids from school. We are focused on our work when our spouse calls to say something came up and they need us to pick up the kids. If we are already depleted and overwhelmed, we might say okay, hang up, immediately go back to what we were doing, then forget to leave on time to pick up the kids.
- We have a recurrent physical therapy appointment on Tuesdays at 10 a.m., but our therapist changes it to Monday just for next week. We are more likely to miss the appointment at that unusual time.
- We have a well-rehearsed morning routine: We make breakfast, make lunches for the kids, pack our work bag, pack a couple of snacks for ourselves, and take our vitamins. If we pick up a phone call as we finish packing our work bag, we might take our vitamins and leave without packing our snacks. (Airplane crashes have occurred when pilots were interrupted while performing their routine sequence of preflight tasks, because after the interruption, they skipped to the next task, not realizing the interrupted task had not been finished.[6])

Prospective memory lapses are also more frequent as we age. As we get older, and when our cognitive resources are drained, we often lose the *content* of the intention: We know there was something we had to do this afternoon, but we cannot remember what.[7] However, our prospective memory does improve in response to cueing and the other strategies described below.

What You Can Try Today

1. *Assess your cognitive needs for today.* Take a couple of minutes to start the day with a rundown of everything you need to get done today: How much will you have to keep in mind? How much working memory and prospective memory support do you feel you need? Do you feel mentally fresh or depleted?

2. *Make checklists.* Make a list, keep it with you, and check it regularly throughout the day. You can use an app on your phone, a bright-colored post-it, or a little notebook. No task is too small to be written down ("Text aunt Sarah happy birthday"). As you accomplish things on the list, do not erase them, just cross them off. This is helpful not only because you can go back and see what you already did (if depleted enough, you might not remember whether you already made that payment), but because it is a small reward to see everything you have already accomplished, and it can give your mood a little boost and increase your feeling of self-efficacy, the feeling that you *can* manage your to-do list.

3. *Take things off your mind and onto paper (or device).* This is key. As soon as you realize there is something you need to do (cancel a subscription, change a plane ticket, dispute a credit card charge), either do it or write it down. You do not need to do everything right away, because this can cause too much shifting, but you need to "offload" it from working memory right away. Add it to your checklist or write yourself a note. I have gotten into the habit of sending myself a text or a voice message if I think of something while I'm driving. I also sometimes text a friend or colleague something like "When I see you, remind me to tell you about an idea I had for Friday's meeting."

Now, if disorganized, notes can become chaotic and unhelpful. Do not have feral notes flying around your desk, bag, and house. Have a centralized system, like the Notes app on your phone or *one* small notepad or daily journal you keep with you at all times, so all your notes are in one place. Having your notes handy is critical, so you don't have to shift off-task for too long. When you remember something in the middle of an activity, you can just easily write it down in the pad or device you have handy. Taking, reviewing, and organizing notes is another helpful use of the "executive" breaks we talked about in Chapter 12.

4. *Set alarms and reminders.* When you are depleted, you might need to set reminders even for things that happen routinely, and certainly every time you schedule a new activity or reschedule an old one. *When* the reminder is set for makes a big difference. One day, while I was working on this chapter, my friend Emily—a talented and versatile psychologist with enviable

cognitive bandwidth—posted this on social media (quoted here with her permission):

> Dropped [son] off at school, drove all the way to work, parked and was walking down the street thinking, "Ok, what do I have on my agenda this morning?" when I realized, "Oh [colorful expression], NOTHING, because [son] has a doctor's appointment this morning." [Facepalm emoji.] An appointment I had on my phone calendar, work paper calendar, work Outlook calendar, and I told [husband] to remind me about this morning. That I knew about for months. I reminded my coworkers yesterday that I needed coverage.

Emily did everything right. She knew her prospective memory might fail her, so she set up several reminders and even "outsourced" to her husband. What happened? Talking about it later, we figured out that she had set reminders to go off both too early and too late. She set reminders for the day before, which worked: She reminded her colleagues that she would need coverage and she emailed her son's teacher to let her know he would be late to school in the morning. She also set a reminder for about an hour before the appointment, but by then she had already dropped her son off and headed to work. (And her lovely husband did forget to remind her.)

If we are depleted or overwhelmed enough, a morning reminder for an afternoon event might not be sufficient, and we might need multiple reminders—one the day before, one the morning of, and one around the time we need to start getting ready or leave for the appointment.

5. Rely on cues. Cues are a depleted person's best friend. Cues significantly decrease cognitive load, because we do not have to keep information in mind—the cue triggers our memory. Cues should be salient and unusual. If you agreed to pick up your neighbor's child on the way to school one morning, put a bright post-it on the garage door the night before. A few times, I have remembered just as I was parking at work that I needed to run an errand on my way home, so I have left an unusual object (like a big floppy hat or an umbrella) on the driver's seat or tied a little bow to the steering wheel. These serve as cues when I get back in the car, tired and distracted at the end of the workday, to remind me of the errand.

6. Piggyback on habits. An effective strategy to help our prospective memory, which is also based on cues, is to link the task to be remembered to an established habit. For example, when you are prescribed a new medication and you are having difficulty remembering to take it, set the pill bottle prominently by your toothbrush or coffee maker. That way you develop the habit of taking the pill before you brush your teeth or drink your coffee.

7. Set intentions. This might feel awkward, but setting an intention to complete an activity and actually visualizing yourself completing it actually works. When you remember that you need to do something important, pause and actually visualize yourself doing it. If you have to leave an event early, for example, but you cannot have an alarm go off, visualize yourself discretely getting up, leaving the room, and heading where you need to go. This increases the chances that you will remember to do it.

8. Rely on supports, but choose wisely. Outsourcing mental tasks is one of the most helpful uses of technology. Smartphones allow us to set customized reminders. GPS allows us to outsource keeping track of distances and routes. Automated bill payments remove due dates from our to-do list. When I travel, I do not even try to remember where I parked my car at the airport, I just take a picture of the area with my phone.

Remember that high-tech is not always best. Do you really need to get notifications from your Roomba or your "smart" refrigerator? Does this decrease your cognitive load or add to it and force you to shift? Times when you are depleted are not the best to try a fancy new organizational app. Low-tech supports work great: Sometimes we remember things better when we write them down, using different colored pens or highlighters. Many medical practices now offer apps, but you can also just ask them to give you a reminder phone call before your next appointment instead. Even if you usually remember to take your medications, if you are depleted or have a particularly busy week, rely on a pill box so you can see whether you already took your medication or not.

9. Talk about it. Normalize relying on aids when your working memory or prospective memory capacity is being taxed. Say, "I'm not going to be able to remember that; let me write it down" or "Can you please send this information to me on an email so I have it in writing?" Include others in your notes review: After a long work meeting or after a family phone call to discuss the plan to help grandma move, send an email with what everyone said they would do, to confirm everyone is on the same page.

References

1. D'Esposito, M., & Postle, B.R. (2015). The cognitive neuroscience of working memory. *Annual Review of Psychology*, 66, 115–142.
2. van Ede, F., & Nobre, A.C. (2023). Turning attention inside out: How working memory serves behavior. *Annual Review of Psychology*, 74, 137–165.
3. Blasiman, R.N., & Was, C.A. (2018). Why is working memory performance unstable? A review of 21 factors. *Europe's Journal of Psychology*, 14(1), 188–231.
4. Brady, T.F., Störmer, V.S., & Alvarez, G.A. (2016). Working memory is not fixed capacity: More active storage capacity for real-world objects than for simple stimuli. *PNAS*, 113(27), 7459–7464.
5. Rummel, J., & Kvavilashvili, L. (2023). Current theories of prospective memory and new directions for theory development. *Nature Reviews Psychology*, 2(1), 40–54.
6. Dismukes, R.K. (2012). Prospective memory in workplace and everyday situations. *Current Directions in Psychological Science*, 21(4), 215–220.
7. Henry, J.D. (2021). Prospective memory impairment in neurological disorders: Implications and management. *Nature Reviews Neurology*, 17, 297–307.

15 Minimizing Multitasking

Many of the most depleting aspects of modern life, most notably the ubiquitous presence of technology and our rushing from one activity to the next, are depleting because they result in our brains constantly shifting from one mental activity to another, back and forth all day long. In work settings, for example, studies have found that we are distracted or interrupted on average about once every two or three minutes, and that we do not get a full hour of uninterrupted work in a day.[1] What happens in our brains and bodies when we go through our days like this?

The Costs of Constant Shifting

The ability to flexibly shift our attention back and forth between two tasks or topics is one aspect of our overall cognitive flexibility, and it is one of the most important cognitive functions underlying our ability to adapt. If we are reading a book, and there is a knock on the door, a loud crash outside, or an alarm on our phone goes off, we appropriately shift our attention to see what is going on. After figuring out what that was—a delivery, a landscaping truck loading heavy equipment, time to take our medication—we can return to our book. Being able to interrupt and resume a task is an incredibly adaptive feature of our brain. However, the idea of "multitasking," or doing more than one thing at a time, does not rely on our ability to flexibly refocus our attention after an interruption, but on repeated, ongoing, and quick shifting back and forth between two tasks, like listening in on a work call while we fill out paperwork or paying bills while watching TV.

Shifting is also a cognitively expensive feature of our brain.[2] Every time we switch our attention, we have to disengage from the first task and re-orient to the new task, which takes time and cognitive effort. The same is true when we return to the first task: We have to disengage from the second task and re-orient to the first one. When shifting back and forth between two activities, we are exerting the normal cognitive effort related to each task, plus additional, precious mental resources spent every time we shift from one task to another. Remember that our attentional control systems are supposed to be allocating resources to the goal at hand: When we keep

DOI: 10.4324/9781003409311-19

switching between tasks, there are multiple goals. We create conflict and interference, confuse our attentional networks, and waste cognitive resources. Switching requires more neural activation than staying on task, so our brains literally have to work harder.

There is also a cost to shifting in terms of time and accuracy. When completing two tasks simultaneously instead of sequentially, our performance slows down and it takes us longer to complete both tasks. If we are reading a book chapter and texting a friend, we take longer than if we read the chapter without interruption and texted a friend without interruption.[3] We are also more likely to make mistakes.

Is multitasking always bad? No. The costs of multitasking are due to interference between the neural activity required for each task, so with extended practice, as a task becomes more automatic, there is decreased neural activity required and interference is reduced.[4] This means we can, to some extent, do two things at a time when one of them is relatively automatic and requires little executive control or intentional thought, and especially when the activities use different cognitive "channels." We can easily talk to a friend on the phone while we go on a walk—one task is primarily motor and relatively automatic; the other one is cognitive and verbal. We can probably listen to an audiobook or podcast while we clean the kitchen and do the laundry. We might be able to cook a routine meal while we help our child with homework at the kitchen table, but we might make mistakes if we are actually trying to follow a new recipe.

Multitasking is often inefficient when we engage in two tasks that are effortful, require intentional control, or use the same "channel." Attending an online meeting and shopping online, for example, are both cognitive tasks involving processing information through verbal and visual channels, working memory, reasoning, and decision-making. Moreover, our ability to shift can change based on how cognitively depleted we are, and we are poor judges of our multitasking abilities, tending to overestimate how well we can multitask.[2]

The "Multitasking" Brain

Shifting is not just costly, it is stressful: When we switch attention repeatedly and rapidly, our heart rate and blood pressure increase, our immune response weakens, and we experience higher levels of stress, anxiety, and burnout.[5] People who are interrupted more while trying to focus on a task experience more stress, frustration, and time pressure, and have to exert more perceived effort (and this is based on experiments where people are interrupted for a short period of time, like 20 minutes, not throughout an entire day).[1]

We also remember less information: If we multitask while information is presented (say, during a lecture or while reading a book like this one), we remember less content.[3] We might be able to remember the most important points, but lose details.[6] And like many of the other things we do to try to save time, multitasking does not always help: It decreases our efficiency because we have to go back and correct mistakes or look for information that we missed while multitasking.[7] If focusing on one thing at a time feels slow, think of all the time you might be currently spending making up for mistakes or incomplete work you did while dividing your attention, shifting between multiple tasks, or managing multiple distractions.

Another interesting consequence of multitasking is that it might make us more prone to distraction. After a period of time when we have to shift because of multiple external interruptions, for example, we self-interrupt more.[1] In other words, after we have been shifting repeatedly for some time because of external distractions, we start internally distracting ourselves—we self-initiate shifting.

What You Can Try Today

1. Eliminate shifting triggers in your environment. Even if you are working on a short task, put a "Do not disturb" sign on your door. Set your phone on "Do not disturb" mode (you can customize your phone setting so that only texts and calls from specific contacts continue to come through if you need to). Even better, put your phone in a drawer so it is out of sight and you are not tempted to check social media.

2. Set boundaries. Check social media and read/watch news only at pre-determined times during the day, and only for a pre-determined period of time. You can tie these activities to other time-limited events—for example, you only check social media or news apps while having your coffee in the morning, or you only listen to one news show during your drive home from work. If possible, turn your office ringer off and listen to messages in your voicemail only at the beginning and the end of the workday (the timing and frequency will, of course, depend on your specific circumstances). Check your email only on the hour, then actually close the app or browser window for the next hour.

3. Delete notifications. Just like reminders and cues are a depleted brain's best friends, notifications are the enemy. Notifications serve no purpose other than capture your attention and lure you away from what you are doing. You do *not* need to know every time you have a new post, comment, text, or email waiting, or every time "breaking news" are posted to a news app. Only keep the absolutely essential notifications active, like new communications from your close family members. Silence everything else. You can check new activity when you intentionally open the app.

4. Mark transitions. Do not go from one thing to the next throughout the day without marking transitions. When you plan your day, leave some time, even if it is just a few minutes, in between activities. When you finish one task, pause to think about what comes next, and do something that signals to your brain you are shifting to something else. Do one of your restorative pause activities from Chapter 12: Stand up, stretch, take a few deep breaths while looking out a window. Pull out your checklist, and cross the task you finished off, as a way to concretely mark its end and the transition. You might even say out loud, "Okay, done with that. Next, we're going to…" You need to switch from one "setting" to the next.

Do the same for interruptions: If the phone rings while you are writing, finish the sentence or complete the thought you are writing before disengaging your attention. Leave a visible "bookmark" or write a few cue words for what you will write when you come back to it after the call. Take a couple of breaths before picking up the phone. Do not mindlessly reach for the receiver, and certainly do not continue to type while you're mindlessly saying "Uh-huh" to the person on the phone. You are setting yourself up for not remembering or misremembering information, missing things that are said, and saying things without thinking them through. After you hang up, immediately offload anything you need from your working memory (do you need to add something to your notes or to your schedule based on the call?), then again take a couple of deep breaths and pause to re-orient yourself to what you were writing.

5. Outsource to technology. There are apps that you can use to block access to certain apps and websites, like social media apps, based on a customized schedule. Some of these apps work on your phone, some on your computer, and some on both; some are free and others are not. Studies have shown that those of us with poor self-regulation skills, who have trouble stopping ourselves from multitasking by opening apps or websites, benefit from these blockers, because they allow us to outsource the work of self-regulating our behavior. However, people with good self-regulation skills, who are able to do this independently, can do worse with these blockers, because the blockers take away their opportunity to take self-initiated, spontaneous breaks. Keep in mind that even if you are typically a good self-regulator, if you are currently cognitively depleted and overwhelmed, you might temporarily benefit from this kind of software.

6. Outsource to other people. Remember that multitasking is not only costly, but reduces your memory for the content of the activities. Use the buddy system: Bring a friend to an important medical appointment or informational session so they can take notes for you while you focus on processing the information and asking questions. For example, my friend Kristen and I had a system for our children's school functions that we kept up through their high school graduations: I took pictures and video during her daughter's performances and functions, and she took pictures and videos during my son's. This way, the other could be fully present taking in the event, making memories without interference.

7. Multitask wisely. Remember the better ways to multitask. Talk to a friend on the phone while you fold and put away the laundry. Listen to a podcast while you cook and store a few meals for the week. Go over your schedule for the day while you eat breakfast. However, be aware that even when you *can* do more than one thing at a time, you will still be giving up the benefits of completing an activity mindfully, with undivided attention. Mindful laundry might not sound like something to be invested in, but eating is a good example of an activity that has been shown to have important health benefits when performed mindfully (the website for the School of Public Health at Harvard University has good information about this—you can search for "mindful eating" at www.hsph.harvard.edu).

8. Talk about it. Normalize mono-tasking. When your children excitedly burst into your room to tell you something, say, "Hold on a second, let me shift my attention." Pause what you are watching on the TV, turn the volume of your music off, put a bookmark in your book. If somebody comes to your office while you are writing an email, say, "Give me one minute to finish this so you can have my undivided attention." At work, raise the possibility of uninterrupted periods of time, when people do not check email, message each other, or interrupt one another, so you can all focus on tasks without needing to shift.

References

1. Mark, G. (2015). *Multitasking in the digital age*. Springer.
2. Madore, K.O., & Wagner, A.D. (2019). Multicosts of multitasking. *Cerebrum*, cer-04-19.
3. Jamet, E., Gonthier, C., Cojean, S., Colliot, T., & Erhel, S. (2020). Does multi-tasking in the classroom affect learning outcomes? A naturalistic study. *Computers in Human Behavior*, 106, Article 106264.
4. Garner, K.G., & Dux, P.E. (2023). Knowledge generalization and the costs of multitasking. *Nature Reviews Neuroscience*, 24, 98–112.
5. Mark, G., (2023). *Attention span: Find focus, fight distraction. A groundbreaking way to restore balance, happiness, and productivity*. Hanover Square Press.
6. Middlebrooks, C.D., Kerr, T., & Castel, A.D. (2017). Selectively distracted: Divided attention and memory for important information. *Psychological Science*, 28 (8), 1103–1115.
7. May, K.E., & Elder, A.D. (2018). Efficient, helpful, or distracting? A literature review of media multitasking in relation to academic performance. *International Journal of Educational Technology in Higher Education*, 15, Article 13.

16 Containing Worry

How often do you catch yourself replaying in your mind past mistakes and things that went wrong? How often do you find yourself sidetracked by thoughts about what might have happened if you had or had not done something? By thoughts about your cognitive changes, health problems, or family problems? About things you have to do but you really do not want to?

So far, we have largely focused on external demands competing for our cognitive resources. But have you ever come home from the grocery store and caught yourself sticking chicken in a drawer? You are not multitasking, and no one is distracting you, so clearly something is wrong with your brain. Not at all. Many glitches are caused by *internal* distractions, and a primary culprit are our ruminative thoughts.

A Busy Brain: Worry and Rumination

Worry and rumination refer to forms of perseverative cognition—negative, repetitive, and unproductive thoughts.[1] Some researchers make a distinction between them, with rumination referring to thoughts about past events, like failures and losses, more commonly (but not exclusively) seen in depression, and worry referring to thoughts about the future, such as things that might possibly go wrong, more commonly (but not exclusively) associated with anxiety.[2] In this chapter, I will use the term *rumination* to refer to both kinds of perseverative thoughts.

The presence of rumination is common in many mental health conditions, including depression, anxiety, trauma, and eating disorders, and it is associated with intense preoccupation with something—a traumatic experience, how terrible our life is, how unhappy we are with our bodies, or how our pain will never go away. In fact, a tendency to ruminate is a risk factor for the development of psychological conditions in general. At the same time, you might remember from Chapter 5 that in conditions like depression and anxiety, people develop a tendency to pay attention to and fixate on negative information, which can make it even more difficult to disengage from the negative thoughts.[3]

Rumination is associated with a higher risk for poor health outcomes. Those of us who ruminate are at higher risk for increased cortisol levels,

DOI: 10.4324/9781003409311-20

impaired ability to recover from the cardiovascular effects of stress (e.g., our blood pressure might remain elevated longer), the development of cardio-vascular disease over time, poorer sleep patterns (it can take us longer to fall asleep, we sleep less, and our sleep is of poorer quality), substance abuse, binge-eating, engaging in self-harm behavior, and having a pessimistic out-look on life.[2], [3], [4], [5] We are also less likely to engage in or respond to pleasant activities, and less likely to engage in active problem-solving.

The Costs of Constantly Running the Worry "App"

When we are ruminating, our brains have to constantly inhibit those perse-verative thoughts so we can focus on the tasks at hand. Because of this, we are left with fewer cognitive resources for the task itself, and to inhibit other distractions and sources of interference.[1] As we mentioned in Chapter 14, rumination also decreases our working memory capacity, and it interferes with our ability to shift our attention.[5] If we are trying to go about our day but we are having persistent, uncontrollable thoughts like "This is never going to work," "I can't do it," and "Nobody ever helps me," it is as though we are constantly multitasking: Our attention has to repeatedly shift back and forth between those thoughts and whatever it is we are trying to do. This is true whether our perseverative thoughts are of a depressive ("I'm all alone"), anxious ("What is going to happen?") or angry ("I have had enough!") variety. [6] Regardless of the "flavor," rumination is a constant app running in our brains, taking up cognitive resources and draining our battery.

Interestingly, the relationship between rumination and executive function seems to be bidirectional: Rumination decreases our executive resources, and when our executive functioning is weak—for example, because we are deple-ted due to sleep deprivation, chronic stress, or illness—we are more likely to ruminate, largely because of our inability to inhibit negative thoughts.

What does this look like in the brain? When we ruminate, there is activation of components of the emotional network, like the amygdala, activation in compo-nents of the emotion regulation network, like areas of the prefrontal cortex, and also increased activation of the default mode network (DMN).[3], [7] You might remember that the DMN is active when the brain is idle, not actively engaged in a task, but also when we are engaged in self-referential processing. This reflects a key component of rumination: We are self-preoccupied. We are less focused on what we are doing, or what is happening around us, because our attention is directed inwards, focused on negative thoughts about ourselves and our lives.

Not only that, but in people who are more prone to ruminate, activity in the DMN does not decrease in the normal way when the person has to engage in a cognitive task, reflecting that the brain has a difficult time disengaging from the ruminative, self-preoccupied thoughts to perform a goal-directed mental activity.[6]

The bottom line is that inhibition is expensive and drains our cognitive resources. Whenever we are inhibiting something, our prefrontal cortex is hard at work. This is true whether we are inhibiting ruminative thoughts, inhibiting our emotions as we will see in Chapter 17, inhibiting a behavior, like eating unhealthy food or relapsing on alcohol, or inhibiting a memory, like intrusive images from an upsetting interaction or a traumatic experience. Even when healthy and rested, suppressing emotions, thoughts, memories, and behaviors is effortful; when we are depleted, it might be impossible. As we will see below, a way to free up cognitive resources is to eliminate the need to inhibit or suppress, by getting the thoughts, emotions, or memories out of our brain.

What You Can Try Today

1. Check how it is going "upstairs." Again, the first step is awareness. Start your day with a check-in: Is your mind quiet? Busy? Is there a lot of chatter? What are the main stressors on your mind right now? Any dominant emotions swirling around your mind and your body? What emotions, thoughts, memories, or impulses are demanding you attention?

2. Get it out. Whenever possible, the most straightforward way to decrease the need to inhibit your ruminative thoughts is to take them off your mind. Journal for a few minutes, so you pour your thoughts out onto the paper. Actually articulate your thoughts, your fears, your concerns, and your hopes. Instead of fighting your thoughts, pretending they are not there, or ignoring them, acknowledge them, and give them a place to exist, on the paper instead of your working memory "notepad." Throughout the day, if you find yourself preoccupied with similar or additional thoughts, similarly write them down. Acknowledge them, accept them, get them out. You can also record voice messages on your phone, as if you were talking to someone else. Remember, it only takes a couple of minutes to feel heard.

3. Get concrete. We will talk more about transforming our thoughts in Chapter 18, but here is one quick strategy. One of the characteristics of ruminative thoughts that makes them unproductive is that they tend to be abstract, like "I'm going to fail" or "Nobody cares." Try to make your thoughts more concrete and figure out what exactly you are telling yourself. If your ruminative thoughts sound like "Nobody helps me," ask yourself concrete questions, meaning "Who," "What," "When," "Where," and "How" questions: "Who helped me the last time I needed help?" "Who have I asked for help so far?" "Whom do I want to ask for help?" "What do I need help with right now?" "What kind of help do I actually want?" "What specifically should I ask for?" "When is the best time today to reach out to [person]?" and so on. You might recognize this from Chapter 4 as a shift towards "approach" coping, a healthier, problem-focused approach that can lead to positive action and solutions.

4. Schedule worry time. This really works. If there is something you are worried about, and you find yourself getting distracted by your thoughts about it, make an appointment with your worry. Decide at what time, where, and for how long you will worry: For example, you will worry about your daughter's difficulties adjusting to her freshman year of college this afternoon from 4:30 to 4:50 p.m. on the couch in the living room. When the time comes, actually worry. This is not problem-solving time; this is worry time. You might notice that when you give your worry your full, undivided attention, it takes less time than you thought to start feeling that the worry is receding.

5. *Distract yourself.* For mild rumination, especially in situations that are outside of your control, distraction can help you break the cycle. If you are truly stuck in your thoughts, take a break to go on a quick walk, watch a couple of fun videos online, play a couple of your favorite songs, and change your environment. Remember that when we ruminate, we fixate on negative thoughts and experiences. Try to break that pattern by intentionally seeking positive experiences: Look for something beautiful, funny, interesting, or awe-inspiring. A short, pleasant activity can help short-circuit mild rumination and prevent it from snowballing.

6. *Return to mindfulness.* Mindfulness is one of the most effective antidotes to rumination. You do not push your ruminative thoughts away; you do not fight them or argue with them. You acknowledge them, but you do not engage with them. You can look up audios of mindfulness meditations for rumination, or simply spend a few minutes taking slow breaths and engaging in the common practices of imagining your repetitive thoughts as leaves on a stream or clouds passing by on a windy day. Observe your thoughts appear, acknowledge them with a nod, and visualize them going by.

7. *Do not ruminate about rumination.* One common experience is that of *secondary emotions*: We feel sad about our anxiety, ashamed about our depression, guilty about our anger. Similarly, we might notice that we start ruminating about our rumination: "Here I go again. What is wrong with me? Get a grip. I have stuff to do! I need this to stop. This is never going to get better. I'm so out of control." And on and on. Remember that some degree of rumination is normal, and that when you are depleted, your brain's attempts at keeping you alive and well can become derailed. Anxiety and worry are ultimately the product of your brain trying to protect you.

8. *Take action.* There are times when we cannot avoid the need to inhibit, such as when we have to suppress our fear, anger, or pain in front of small children. But many other times, acting on the ruminative thought, rather than inhibiting it, is appropriate. Is this something that can actually be addressed or problem-solved? If you keep ruminating about a comment somebody made, reach out to clarify what they meant. If it is a situation you do not have control over, perhaps you can at least reduce the uncertainty around it. Let's say you are ruminating about the result of a medical test or about a job interview: Call the medical office to ask when you should expect the results to be available; call the hiring office to ask when you should expect to hear back. Reducing uncertainty, even if we cannot eliminate it, can help contain rumination.

9. Talk about it. Normalize rumination. Talk to others about it. If appropriate, share your ruminative thoughts out loud. Ask others to carry your worry with you. Say, "Hey, I know we already talked about this problem and came up with a plan, but I'm still distracted thinking about it," "I know the doctor said it's nothing, but I'm still worried about it," or, "I'm stuck on a comment you made earlier today." Remember the hill slant studies from Chapter 10? Sharing our concerns with others helps change our perceptions. They can also help us think about the situation in new, healthier ways, identify resources, and receive practical help.

Sharing our ruminative worry can benefit the relationship itself, because when we are preoccupied with our thoughts, especially when our worries are about situations involving someone else, our worry can become this obstacle sitting between us and the other person. Let's say we are worried about our spouse's lack of motivation at work. We have started ruminating about it, worrying excessively, and nagging them. One way to visualize this is that each of us is sitting on opposite sides of our ruminative worry, so neither of us can quite see what the other sees. When we share our thoughts, however, we bring the other person to our side of the problem. It is like saying, "This is what I'm seeing. What do you see?" Sharing our worries can bring us together instead of standing between us.

References

1. Beckwé, M., Deroost, N., Koster, E.H.W., De Lissnyder, E., & De Raedt, R. (2014). Worrying and rumination are both associated with reduced cognitive control. *Psychological Research*, 78, 651–660.
2. Clancy, G., Prestwich, A., Caperon, L., Tsipa, A., & O'Connor, D.B. (2020). The association between worry and rumination with sleep in non-clinical populations: A systematic review and meta-analysis. *Health Psychology Review*, 14(4), 427–448.
3. Watkins, E.R., & Roberts, H. (2020). Reflecting on rumination: Consequences, causes, mechanisms, and treatment of rumination. *Behaviour Research and Therapy*, 127, Article 103573.
4. Murray, R.J., Apazoglou, K., Celen, Z., Dayer, A., Aubry, J-M., Van De Ville, D., Vuilleumier, P., & Piguet, C. (2021). Maladaptive emotion regulation traits predict altered corticolimbic recovery from psychosocial stress. *Journal of Affective Disorders*, 280, 54–63.
5. Yang, B.S., Cao, S., Shields, G.S., Teng, Z., & Liu, Y. (2017). The relationships between rumination and core executive functions: A meta-analysis. *Depression and Anxiety*, 34, 37–50.
6. du Pont, A., Rhee, S.H., Corley, R.P., Hewitt, J.K., & Friedman, N.P. (2019). Rumination and executive functions: Understanding cognitive vulnerability for psychopathology. *Journal of Affective Disorders*, 256, 550–559.
7. Vrshek-Schallhorn, S., Velkoff, E.A., & Zinbarg, R.E. (2019). Trait rumination and response to negative evaluative lab-induced stress: Neuroendocrine, affective, and cognitive outcomes. *Cognition and Emotion*, 33, 466–479.

17 Befriending Our Emotions

One of the biggest misconceptions about the human brain is that "emotion" and "cognition" are distinct brain processes. We experience emotions and cognitive functions as separate, but in the brain, cognition and emotion are intertwined. [1], [2] The networks responsible for cognitive functions like attention, memory, and executive functions overlap with those that process emotional experience and regulate emotions. Think of emotion and cognition as two different colors of Play-Doh mixed together: You can tell both colors apart, but you are not able to neatly separate them. Because of this, we cannot decrease cognitive overload without paying attention to our emotions and our attempts at regulating them. The next two chapters focus on how to do that.

Emotion Regulation

Emotions have multiple components:

- a subjective or experiential component—what it actually *feels* like to be happy, scared, angry, or sad;
- a physiological component, like the palpitations of fear, the pit in our stomach of anxiety, the flushed face of embarrassment;
- a cognitive component—thoughts like "I can't believe he said that to me!" "I want to get out of here," and "I made a fool out of myself"; and
- a behavioral component—for example, fear might make us want to flee, anger might make us want to say something hurtful, and happiness might make us want to hug someone.

Attempts at emotion regulation vary in their nature, the component of emotion they address, and their outcomes.[3], [4] Often we try to down-regulate or decrease emotion, as when we try to calm down our anxiety. Other times we try to up-regulate emotion or turn the emotional volume up, as when we try to get excited enough to attend a social function, or when we indulge in watching sad movies to get "a good cry."

We can use emotion regulation to try to prevent emotion by selecting what situations we go into and which ones we avoid based on their likely

DOI: 10.4324/9781003409311-21

emotional consequences. If we get invited to a large Thanksgiving dinner at a relative's house, but we know their spouse is a heavy drinker who enjoys loudly arguing about political opinions we find jarring, we might choose to avoid that situation (and the emotions we would experience) by skipping it and instead having a quiet dinner with a couple of close friends.

If we find ourselves already in a situation that triggers an emotional reaction, we can attempt to modify the situation or problem-solve. For example, if we do find ourselves at dinner with the loud, argumentative host, we can try to gently redirect the conversation, more assertively tell them to please tone it down, or start a party game to defuse the tension.

If we are in a situation that triggers an emotional reaction and that we cannot quite change, we can choose to focus our attention on different aspects of the situation to help regulate our emotions. In the example above, we could choose to distract ourselves by focusing our attention on our cousin's new baby while the host continues their rant.

We can also change how we think about an emotional situation and reframe our perception of it. For example, instead of dismissing our vocal host as rude and unpleasant, we can see them as someone who is concerned about his family's future, frustrated and isolated since his retirement, and turning to alcohol for relief, which can transform our frustration into empathy. This kind of *cognitive reappraisal* is an important emotional regulation strategy deserving of its own chapter, and we will focus on it on Chapter 18.

Finally, we can attempt to regulate our emotional response instead of the situation. For example, we can try to suppress the outward expression of emotion: We can try not to cry when our feelings are hurt, not to smile when we find someone's mishap amusing, speak softly when we are angry, and straighten our spine and lift our chin when we feel embarrassed. We can also try to change the physiological components of emotion—for example, by taking deep breaths (or using substances) to try to calm down when we are nervous.

Many problematic behaviors are actually attempts at emotion regulation. Substance abuse can begin as an attempt to escape negative emotions like anxiety, depression, or shame. Procrastination is an emotion regulation strategy based on avoidance: We often procrastinate tasks that are anxiety-provoking, like paying our bills when we are short on cash. Staying on our phones until we are literally nodding off can be another way we avoid uncomfortable emotions like anxiety by preventing us from thinking.

Our Brains on Emotion

The inextricable connection between emotion and cognition means our cognitive and emotional health are equally inextricably connected. As we reviewed in Chapter 5, most psychological conditions have some impact on our cognitive functioning. Cognitive rehabilitation programs for people with neurocognitive disorders—for example, due to a traumatic brain injury—recognize this bidirectional relationship and incorporate emotional regulation training as part of the intervention.[5] This includes developing awareness of the physiological, experiential, and cognitive aspects of emotions, understanding the triggers and consequences of our emotional responses, and training on emotion regulation strategies.

In general, some emotion regulation strategies, such as problem-solving and reframing emotional situations, tend to be more helpful and lead to better outcomes. Suppressing the expression of emotions is effortful, relying on inhibitory systems, and it does not usually decrease our negative emotional experience.[3] In fact, in can amplify our physiological response (and yet think about how often we suppress the expression of our emotions, as parents, caregivers, employees, etc.). Adults who engage in suppression routinely have lower levels of well-being, more symptoms of psychopathology, and less satisfaction with their relationships.[4]

Importantly, there are individual and cultural differences. Some people naturally feel emotions less intensely or are less emotionally expressive. What is unhelpful is when we feel emotion intensely and we try to keep a "poker face"; it is in those cases that we end up feeling worse. Similarly, individuals who suppress their emotional expression are rated by others as less desirable social partners, but this is only the case in Western cultures, not in Eastern cultures with different social norms for emotional expressivity.[3]

That being said, every strategy can be adaptive under certain circumstances. For example, while relying on distraction and avoidance routinely can lead to problems, distraction does successfully decrease negative emotion and reduce amygdala activation, and it can be effective when we need to change our emotional experience quickly, in response to intense stimuli and in situations we cannot escape, like watching a light-hearted movie during a painful medical procedure.[3]

What You Can Try Today

1. *Start with awareness.* Check your emotional landscape today. What are the most salient emotions you are experiencing? Is the overall tone positive and pleasant, or negative and uncomfortable? How intense are they? Mild and lingering in the background, or more intense and disruptive? The exercises in Chapter 19 can help by increasing awareness of your mind and body, but you can also do this while having your morning coffee or taking a shower.

2. *Give yourself permission to avoid in healthy ways.* While it is unhealthy to rely on avoidance as a long-term emotional regulation strategy, it is okay to avoid upsetting situations under some circumstances, especially when depleted. You can avoid contact with people whose presence drains you emotionally—it is okay not to pick up the phone or not to stop to chit-chat with acquaintances during school pick-up. Bow out of stressful situations

like large gatherings and pass on activities that are too demanding for your cognitive resources at this time, like volunteering to run the school book fair.

3. Learn how to use distraction to your advantage. Similarly, while you do not want to rely on distraction as a long-term strategy, there is room for it in our emotion regulation repertoire. It can be healthy to distract ourselves when faced with a time-limited uncontrollable stressor like waiting for the results of a medical exam or waiting for our loved one to come out of surgery.

4. Problem-solve. Solution-focused approaches to coping with stress and negative emotions lead to better outcomes and better overall health, but can seem effortful when we are depleted, making avoidance more tempting. There is a commonly used heuristic for problem-solving that, while simple, is actually quite helpful. When you find yourself in a stressful situation that you need to address or solve, ask yourself, "Is this important?" and "Is this urgent?"

Important and urgent stressors need your dedicated time and energy and take precedence over other stressors. For example, your mother fell ill unexpectedly, is getting discharged from the hospital in a couple of days, and you need to find home care for her. This is the moment to set everything else aside and focus your time and energy on problem-solving for this situation. *Very* few problems fall in this category.

When a stressor is important but not urgent, consider your options and allocate a set amount of time to the problem. For example, if your children's summer camp gets canceled and you need to come up with a different plan in the next month, choose a few options you will explore, pick a few people you will talk to, and set a deadline for when you will make a decision. Prioritize this stressor, but keep it contained.

If the stressor is not important but feels urgent, do not spend precious cognitive resources on it. You are running late for soccer practice and the coach calls to ask you to pick up snacks. You are tired and overwhelmed already, and you feel like this will push you over the edge. But does it really matter what you do, what snack you get, or even if you are a few minutes late for practice? If it does not, make a quick decision about what snacks you will get and where, and move on. When you are depleted and rushing, urgent things might feel important, but they rarely are.

Finally, let go of anything that is neither important nor urgent. When you are already depleted, you do not need to make time for these. Give yourself a break, let someone else take care of it, and do not add it to your to-do list.

5. Express your emotions. We often find ourselves in situations when we have to hide our frustration, anger, fear, or distress. Remember that suppressing the expression of our emotions can increase our physiological response and make us feel worse. So make sure you find the time and space to express your emotions when you can, however you can, and try to make room for the different components—call a friend to vent your emotions, journal so you can articulate your thoughts, exercise or meditate to release physical tension.

6. *Induce positive emotion.* This is critical. Remember that when we are anxious, depressed, or stressed, we tend to focus on the negative, so we have to very intentionally create positive, pleasant experiences, especially when depleted and overwhelmed. Small pleasant events result in brief releases of dopamine that we experience as positive affect.[6] These natural dopamine surges are transient, so we need to repeatedly seek moments of joy, fun, and contentment through the day. (I have a separate social media account dedicated to positive content solely for this purpose. I do not use it to interact with anyone, just to follow accounts that post nature photographs, funny animal videos, art, and comedy. I check it for a few minutes at a time when I need a positive "boost.")

Cognitively, positive emotions broaden our attention, make us more creative, and lead us to be more open to information and consider a wider range of possible actions.[7] There is also some evidence that positive emotions might, under certain circumstances, help us recover from the cardiovascular impact of negative emotions and stress.[8] (To learn more about research on the effects of positive emotions on cognition, look up the talk *Positive emotions open our mind*, by Barbara Fredrickson.)

There are many strategies that increase positive emotions.[9] Imagine positive outcomes before an event (e.g., visualize yourself doing a great job with the presentation you are giving later today). Journal about positive topics, like things you are grateful for, but also your dream scenarios. Play music you find beautiful or that makes you feel energized. Celebrate even small successes—get yourself a little treat, or text a friend to share the good news. Perform an act of kindness for someone else. Take a short walk with the sole purpose of noticing as many positive things as you can; you can take a picture of something pleasant, like a beautiful tree, and share it with a friend. Reminisce about cherished memories by invoking vivid mental imagery.

7. *Outsource your emotion regulation to others.* This is one of my favorite facts about human brains: When we are experiencing negative emotions but a loved one is present, we can outsource our emotional regulation to them. [10] More specifically, when we are in a negative emotional state, many areas of our brain involved in the different components of the emotional response are active.[11] If a stranger holds our hand, there is some decreased activation in brain areas involved in the regulation of our emotional response. But when it is our spouse holding our hand, there are even greater decreases, including in areas involved in effortful emotion regulation, and the effect is stronger the better the quality of our relationship. In other words, the presence of someone we are emotionally close to diminishes our negative emotional response *without* the need to activate the brain regions associated with effortful emotional control. To learn more about this fascinating topic, I encourage you to look up the TED talk *Why we hold hands*, by Jim Coan.

8. *Talk about it.* Normalize emotions and emotional regulation in daily life. Share the emotional tone(s) of your day with others when you come

home at the end of the day. Say to your children, "I had a rough day today, I'm a little sad. What do you think might help me feel better?" Model the language of emotions and healthy emotion regulation. Tell your coworker, "I'm frustrated about the discussion at that meeting." By acknowledging and naming your emotional reactions, you are avoiding the suppression of every emotional component. Seek ways to express your emotions as you experience them, in ways that are appropriate for the social context.

References

1. Feldman Barrett, L. (2017). *How emotions are made: The secret life of the brain.* Mariner Books.
2. Davies, A., Rogers, J.M., Baker, K., Li, L., Llerena, J., das Nair, R., & Wong, D. (2023). Combined cognitive and psychological interventions improve meaningful outcomes after acquired brain injury: A systematic review and meta-analysis. *Neuropsychology Review*, 2023.
3. McRae, K., & Gross, J.J. (2020). Emotion regulation. *Emotion*, 20(1), 1–9.
4. McRae, K. (2016). Cognitive emotion regulation: A review of theory and scientific findings. *Current Opinion in Behavioral Sciences*, 10, 119–124.
5. Cantor, J., Ashman, T., Dams-O'Connor, K., Dijkers, M.P., Godon, W., Pielmanm, L., Tsaousides, T., Allen, H., Nguyen, M., & Oswald, J. (2014). Evaluation of the Short-Term Executive Plus intervention for executive dysfunction after traumatic brain injury: A randomized controlled trial with minimization. *Archives of Physical Medicine and Rehabilitation*, 95, 1–9.
6. Sterling, P. (2020). *What is health? Allostasis and the evolution of human design.* MIT Press.
7. Fredrickson, B.L. (2001). The role of positive emotions in positive psychology: The broaden-and-build theory of positive emotions. *American Psychologist*, 56(3), 218–226.
8. Behnke, M., Pietruch, M., Chwiłkowska, P., Wessel, E., Kaczmarek, L.D., Assink, M., & Gross, J.J. (2023). The undoing effect of positive emotions: A meta-analytic review. *Emotion Review*, 15(1), 45–62.
9. Quoidbach, J., Mikolajczak, M., & Gross, J.J. (2015). Positive interventions: An emotion regulation perspective. *Psychological Bulletin*, 141(3), 655–693.
10. Reeck, C., Ames, D.R., & Ochsner, K.N. (2016). The social regulation of emotion: An integrative, cross-disciplinary model. *Trends in Cognitive Neuroscience*, 20(1), 47–63.
11. Coan, J., & Maresh, E.L. (2014). Social baseline theory and the social regulation of emotion. In: J.J. Gross (Ed.), *Handbook of emotion regulation* (2nd ed., pp. 221–236). Guilford Press.

18 Transforming Our Thoughts

In the beautiful, poignant book *Everything happens for a reason (and other lies I've loved)*, Kate Bowler shares the responses people had to an article she wrote about being diagnosed with stage 4 cancer at age 35, when her son was one year old.[1]

> "It doesn't matter, in the End, whether we are here or 'there.' It's all the same."
> "We can't always get what we want."
> "Everything happens for a reason."

People talk about what she will learn through her suffering. They say that her cancer is all part of God's plan, so she can help people through her writing. That it is a test of her faith. That it is a consequence of her sins.

> "Keep smiling! Your attitude determines your destiny!" says Jane from Idaho, and I am immediately worn out by the tyranny of prescriptive joy.

* * *

"Be grateful for what you will learn from this." "Look on the bright side." "Good vibes only." *Cognitive reappraisal* is a helpful emotion regulation strategy associated with many good health outcomes. It involves changing the way we think about a situation in order to change how we feel.[2] But it is not the tyrannical "prescriptive joy" Dr. Bowler is talking about. What is it, then?

"Think Happy Thoughts"?

Our thoughts do affect how we feel and how we act, so changing how we think about situations does change how we feel. In fact, many effective forms of psychotherapy are developed around this principle. Let's say we hate our job. Cognitive reappraisal might mean, for example, that instead of thinking

DOI: 10.4324/9781003409311-22

of it as "boring," we think of it as easy, not stressful or demanding. Instead of thinking about it as a "dead end," we think about it as a temporary situation that allows us to regroup and figure out what we want to do next, while still getting us out of the house in the morning and providing us with a paycheck.

Reappraisal works. Reappraising negative situations results in decreases in negative emotion, decreases in the physiological component of the emotional response, and better recovery of our cardiovascular system after the stress response.[3], [4] This is important because, as we saw in Chapter 4, chronic stress levels are associated with poorer health, so shortening the physiological component of the stress response can prevent the wide negative impacts of chronic stress.

Consistent with this, people who use reappraisal as a habitual emotion regulation strategy have better health, have less psychopathology, perform better academically, and have better social lives.[2], [3] In fact, we think about reappraisal as a protective factor: While stressors often increase the risk for developing psychological symptoms, this is not the case if we habitually engage in reappraisal in response to the stressors.[5]

When is reappraisal not particularly helpful? When the negative emotion is intense, requires a fast response, and/or is related to a concrete trigger.[3] By "concrete," I mean a trigger that is not open to interpretation: If we encounter a bear during a hike, if rioting erupts while we are walking down the street, or if our airplane encounters severe turbulence, reappraisal is not a helpful strategy to regulate our emotions. In fact, trying to reappraise a dangerous situation can increase activation in the amygdala—our brain might be signaling that this is the time to seek safety, not think positively.[2]

Reappraisal is also not as helpful when the problematic situation is under our control; then, it is best to use a problem-solving approach and actually change the situation. If we hate our job but we have many other options available, it is best to take the steps to switch jobs to one we enjoy. But if we have no other opportunities and we need the income, reappraisal will be effective at improving our emotional experience while we have to stay.

Why Reappraisal Works

Reappraisal is an effortful emotion regulation strategy, requiring engagement of the prefrontal cortex and some language centers (it makes sense, since reappraising can feel like talking to ourselves).[2] It successfully decreases activation in brain regions implicated in emotional responding, such as the amygdala.[3] In contrast, when we engage in unhelpful responses after a stressful situation, like catastrophizing ("This is the worst," "My career is over"), blame, or rumination, there is increased activity in emotion processing circuits, suggesting that we worsen or prolong the negative emotional experience.[4]

Why is reappraisal such an effective and powerful strategy? The answer is likely complex, but there are a couple of important aspects. First, the efficacy of reappraisal reveals one fundamental feature of our emotional lives, which is that emotions are creations: We experience what we believe.[5] Remember that our brains are always predicting what might happen next based on our experiences and learning. We perceive, experience, and react to what we expect will happen. Say we are meeting our partner's parents. If, based on what our partner has said, we expect them to dislike us, we might interpret their facial expressions, tone of voice, and sparse conversation as confirmation that they do. This might make us feel tense and make the entire experience unpleasant. It is our expectations, which shape our perceptions, that shape our emotional experience. Changing our thoughts changes our expectations and thus our experience.

A second reason is that specific, automatic thoughts often reflect broader, deeper *beliefs* about ourselves, others, and the world. If we have a deep-rooted belief, for example, that we are unlikeable, whenever we find ourselves in a social situation, we might have automatic thoughts that this particular person is not enjoying talking to us. At the same time, ongoing, repetitive thoughts can feed deeper beliefs. Thoughts like "I'm bad at algebra," over time and along with others, can snowball into "I'm dumb." We hold these broader beliefs without challenge, to the point that they become part of ourselves, the filters through which we perceive reality. But by repeatedly reframing our thoughts, day to day, into more adaptive ones, we can chip away at those harmful beliefs.

What You Can Try Today

1. Watch your thoughts. Our brains are constantly producing thoughts. Start by becoming aware of them. What are you telling yourself today? "I'm so tired." "I really don't want to do this." "This is too much." "I'm in so much pain." "There's no way I'm working out today." Remember the mindfulness principles we reviewed in Chapter 16. Do not hold on to your thoughts, do not argue with them. Just notice them. Remember the "leaves on a stream" and "clouds passing by" metaphors. Watch them go by without grasping them.

2. Do not believe everything you think. Some thoughts are strokes of genius. Some thoughts are actual art. But most thoughts are simply products of our brain's constant activity, made from scraps of experience. One very helpful strategy is to add "I'm having the thought that" to our thoughts, as a way to acknowledge their non-literal nature and distance ourselves from their content.[6] *"I'm having the thought that* my forgetfulness means I have dementia." *"I'm having the thought that* my wife hated the present I got her." *"I'm having the thought that* nothing will ever help with my arthritis."

3. Challenge your thoughts. Identify and call out the distortions in your thinking. There are many common cognitive distortions, including:

- Black-and-white thinking: *"Nothing* helps with my back pain."
- Overgeneralization: "She *always* criticizes me!"
- Personalization: *"I* ruined it for everyone."
- Should-ing: "It *shouldn't* be this hard to cut back on my drinking."
- Catastrophizing: "Mom didn't pick up the phone—something terrible has clearly happened."

Look up descriptions of common cognitive distortions so you know what to look out for. The websites for Positive Psychology (www.positivepsychol ogy.com) and the University of Pittsburgh Medical Center HealthBeat (http s://share.upmc.com), for example, have helpful material (search for "cognitive distortions").

Next, practice re-writing your thoughts eliminating the distortions. *"Nothing* helps with my back pain" can become "The three treatments I have tried so far reduced my pain but did not completely take it away." "She *always* criticizes me!" can become "The last few times I did something for her, she said thank you and then commented on the things she would have done differently." "This is *impossible"* can become "I really really really don't want to put in the work it would take to get this done."

4. Short-circuit the snowball. Notice when you are jumping from a thought to a broader belief. If your babysitter tells you last minute that she cannot watch your children tonight, it is normal to have thoughts like "This is frustrating," "I wish she had told me sooner," and even "This sucks." But "This is so typical, you can't trust anyone" is a broader, deeper belief that is likely impacting your general perceptions and experiences.

While working on our more specific, automatic thoughts can help us uncover some of our broader, deeply held beliefs, changing those deep-rooted beliefs likely requires working with a therapist to better understand their origins and consequences and learn and practice other skills. For information on some evidence-based cognitive approaches, Steven Hayes's website (www.stevenchayes.com) has online tools and podcasts related to acceptance and commitment therapy (ACT). There is also a self-help work-book by Dr. Hayes based on ACT principles, *Get out of your mind and into your life.* The Beck Institute, founded by the creator of cognitive behavioral therapy (CBT), Aaron Beck, has a website with resources for the public (http s://cares.beckinstitute.org). Many therapists offer services via telehealth, which is a great option for those with limitations due to chronic illness, caregivers, or those in rural areas with few providers available nearby.

5. Reappraise the stressor. Most of the time, we focus on reappraising the stressor. A few ways to do that involve, for example, reframing the stressor as

- less negative—for example, turning "I made a fool out of myself" into "That was not my best but it was probably not as bad as it felt,"
- more common—for example, "Everybody has bad days, people under-stand," or

- transient—for example, "Nobody will be thinking about this tomorrow," and "I don't like this feeling but it will pass."

Interestingly, it might be most helpful to try to reframe the stressor not just as neutral but as positive. If you have the flu and are dreading spending the weekend sick and alone at home, instead of reappraising it neutrally as "It's not that bad" and "It will go by quickly," you might try, "A quiet weekend is the perfect way to recover" and "This rest is exactly what my body needs."

6. *Reappraise the response.* Some situations are objectively, undeniably difficult, even terrible. How do you reappraise your child being diagnosed with a serious medical condition, or losing your job when you are the sole provider for your family?

Reappraisal is not about what has been rightfully called *toxic positivity*, coming up with superficially encouraging and motivating statements that seek to minimize or erase painful experiences.[7] In situations when the stressor itself cannot really be reappraised, we can still reframe our response.

One way to do this is looking out for distortions specifically about control. For example, here are some thoughts you might have after your parent is diagnosed with dementia:

- "I hate to see my mom declining." There is no reappraising this one.
- "I don't want this to be happening." This is fair.
- "There's nothing I can do." We can reappraise this one. Remember that reappraisal actually works in situations we cannot control. For example, one reappraisal that many dementia caregivers embrace is seeing the illness, as devastating as it is, as an opportunity for them to express their love to the person with dementia. You cannot reappraise the dementia, but you can reappraise your response from "There is nothing I can do" to "I will do everything I can to make mom feel loved, safe, and comfortable for as long as possible."

(Remember that for the aspects you *can* control, problem-solving is the best strategy: You can attend webinars to learn about the disease, join a support group, and start exploring options for engaging activities that are safe for people with dementia.)

A second way to reappraise our response is to tie it to our values. To use caregivers as an example again, many caregivers report lower levels of stress and depression because they find *meaning* in their caregiving: Caregiving for their loved one, while extremely challenging, is consistent with their values. Imagine a family member relapses into drug use, and you agree to take in their three young children for a month while they complete residential treatment. You might find yourself having thoughts about the situation itself like "This is not fair," "This shouldn't be my problem," and "I have enough on my plate." It might be difficult to argue with those. However, if you find yourself having thoughts about your response like "I must be an idiot" or "I'm being taken advantage of," you can reappraise those thoughts, if appropriate, as "I believe family comes first, and even if it is extremely difficult, stepping in and caring for these children during a time of need is the right thing to do."

7. *Talk about it.* Normalize having and reframing unhelpful thoughts. Tell your physician, "I'm hearing what you are telling me, but what I'm thinking is that my life is over." Ask others to "check" your thoughts for you. Once I walked into my colleague Kate's office and told her, "I just got this email that is making me feel that I royally messed up and I need you to come look at it and tell me if that is what it says," because I knew I couldn't believe my thoughts in that upsetting moment. We can work on reappraisals with our kids, our partner, and our friends. We can practice with each other and learn from each other how to change our thoughts in ways that help us feel better and act in more adaptive ways.

References

1. Bowler, K. (2018). *Everything happens for a reason (and other lies I've loved)*. Random House.
2. McRae, K. (2016). Cognitive emotion regulation: A review of theory and scientific findings. *Current Opinion in Behavioral Sciences*, 10, 119–124.
3. McRae, K., & Gross, J.J. (2020). Emotion regulation. *Emotion*, 20(1), 1–9.
4. Murray, R.J., Apazoglou, K., Celen, Z., Dayer, A., Aubry, J-M., Van De Ville, D., Vuilleumier, P., & Piguet, C. (2021). Maladaptive emotion regulation traits predict altered corticolimbic recovery from psychosocial stress. *Journal of Affective Disorders*, 280, 54–63; Riepenhausen, A., Wackerhagen, C., Reppmann, Z.C., Deter, H.-C., Kalisch, R., Veer, I.M., & Walter, H. (2022). Positive cognitive reappraisal in stress resilience, mental health, and well-bring: A comprehensive systematic review. *Emotion Review*, 14(4), 310–331.
5. Feldman Barrett, L. (2017). *How emotions are made: The secret life of the brain*. Mariner Books.
6. Hayes, S.C., Levin, M.E., Plumb-Vilardaga, J., Villatte, J.L., & Pistorello, J. (2013). Acceptance and commitment therapy and contextual behavioral science: Examining the progress of a distinctive model of behavioral and cognitive therapy. *Behavior Therapy*, 44(2), 180–198.
7. Reynolds, G. (2022, September 23). Toxic positivity. Anxiety & Depression Association of America. https://adaa.org/learn-from-us/from-the-experts/blog-posts/consumer/toxic-positivity.

19 Back to the Body

In her gorgeous poem "Tincture," Andrea Gibson imagines a soul missing the body after death, a soul wandering the universe missing laughter, touch, voice, and dreams but also hunger, tears, scars, and pain.[1] In our daily lives, we often neglect the body we exist in, ignore its calls for attention, and take its functions and sensations for granted. We also tend to act like the brain is separate from the body. Luckily for us, the fact that they are not means that the body is yet another arena in which to work on our brain's health. In addition to ongoing maintenance of our body by addressing the factors reviewed in Part II, there are specific strategies that, by focusing on the body, reach the brain and improve its functioning.

Body-Centered Approaches

Psychotherapeutic approaches have long recognized that the boundaries between body and mind, between physical and psychological experience, are artificial. Psychological experiences are embodied and grounded in their physical components, and the state of our body inevitably influences our psychological state. Pretending that we can have a healthy mind while neglecting the body is (pardon the expression) like pretending that we can have a peeing section in a pool. Consistent with this, body-centered approaches, especially those focused on the breath and on muscle relaxation, have a long history in psychotherapy and mental health interventions. In the last few decades, interventions based on mindfulness have also received much attention and scientific scrutiny.

At the core of mindfulness approaches is present-moment awareness with an attitude of nonjudgment, acceptance, and curiosity.[2] There is evidence that mindfulness-based therapies, including mindfulness-based stress reduction, mindfulness-based cognitive therapy, and acceptance and commitment therapy lead to lasting increases in positive emotions and decreases in negative affectivity.[2], [3] Mindfulness-based interventions have been shown to reduce stress levels and improve levels of well-being, work performance, sleep quality, and sleep duration.[4] Mindfulness practice can also increase our awareness of our physiological state, like fatigue levels,

DOI: 10.4324/9781003409311-23

and can help us self-regulate our behavior better. For chronic pain patients who tend to avoid body sensations, increased body awareness through mindfulness training can result in improved levels of well-being.[2]

In addition, brief mindfulness training programs (as brief as one session in duration) focused on learning how to bring nonjudgmental awareness to the present moment have been similarly shown to reduce negative emotion and improve mood, although, not surprisingly, the effects of such short interventions seem to fade over time.[5], [6], [7] Brief mindfulness exercises can decrease pain levels, likely by altering autonomic activity, specifically enhancing parasympathetic activation.[5] One-session mindfulness meditation interventions have also been shown to reduce substance use and anxiety and improve markers of cardiovascular health over subsequent days.

How might mindfulness lead to such benefits on our emotional health? It has been proposed that mindfulness enhances attentional control and emotion regulation skills—for example, by decreasing mind-wandering and rumination and increasing our tolerance of negative emotional thoughts, experiences, and arousal.[5] When we are able to sit with our negative thoughts and emotions, we are less likely to engage in unhealthy coping and emotion regulation strategies like suppression or avoidance.

The Body as the Road to the Brain

Engaging in brief mindfulness exercises can improve attention, decrease mind-wandering, decrease attentional lapses, and counteract the mental fatigue caused by multitasking.[4], [6], [7] Effects are more mixed for more complex cognitive functions, but some improvements have been documented in executive functions, problem-solving, and emotion regulation after engaging in brief mindfulness exercises.[7], [8]

More in-depth mindfulness training has been found to result in modest but significant improvements in executive functions including working memory, with older adults in particular benefiting cognitively from the training.[9] Mindfulness training has also been shown to improve executive function, behavioral symptoms, and emotional regulation in individuals with ADHD. [9] Another potential benefit of mindfulness is an increase in self-monitoring.[10] As we have previously discussed, self-monitoring is an executive function that can be affected when we are depleted or overwhelmed. By enhancing self-awareness, mindfulness can improve our ability to accurately assess our current state (e.g., how fatigued or scattered we are) and to detect our mistakes and lapses.

What You Can Try Today

Before we continue, a word of caution: While body-focused approaches like breathing exercises, progressive muscle relaxation, and mindfulness meditation are effective and helpful for many, some adverse effects have been documented.[2] People with chronic respiratory illnesses should consult with their physician before attempting exercises that involve controlled breathing. Similarly, those with chronic pain should consult with their physician before attempting exercises that involve muscle tensing and relaxation. If, after consultation, it is decided that these approaches are safe to try, it is best to attempt them with a trained therapist rather than independently.

In some individuals, high levels of self-focused attention have actually been associated with negative emotional experiences and even worsening symptoms of depression, anxiety, dissociation, and substance abuse; decreased ability to tolerate pain; and increased arousal, emotional intensity, panic, and traumatic flashbacks. For example, any exercises that involve focusing on body sensations for periods of time might be psychologically aversive for those living with chronic pain. The problem seems to arise when the practice is centered on self-focused attention, with not enough emphasis on doing so with an attitude of nonjudgment and non-reactivity. In general, individuals with certain mental health conditions, including panic disorders, dissociative disorders, and trauma-related disorders, who wish to pursue mindfulness training should do so with a therapist, ideally one with experience working with individuals with the relevant condition.

Finally, while I encourage you to spend some time searching for and trying different guided exercises online, you will find a large number of tutorials on platforms like YouTube and SoundCloud and it can be difficult to know which ones to focus on. I recommend starting with videos and playlists by healthcare and mental health organizations. Several college counseling and student wellness centers also have online resources, including guided exercises, that are freely available.

1. Check in with your body at the beginning of the day. You can look up "body scan" guided exercises to see if this is something you might be interested in trying. These tend to be longer, meditation-like exercises with the goal of bringing attention to different parts of your body, although shorter versions are also available. Body scans do not attempt to change the state of your body, but to connect you to it by directing your attention and increasing your awareness of your body—to get you to simply notice it. Some of them incorporate elements of self-compassion, which can be helpful for those living with chronic disease and other physical symptoms.

2. Calm your body. Progressive muscle relaxation (PMR) is a practice that is effective at relaxing the body and quieting the mind. It can increase deep sleep, so it can be helpful to do it before bedtime. It involves tensing different muscles in the body, then releasing the muscle tension. The purpose of doing both steps is to create awareness of what tension and relaxation feel like. There are many scripts available, of varying length, and focusing on a different number of muscle groups.

If PMR feels uncomfortable to you, there is a similar intervention simply called Progressive Relaxation, which instructs you to relax your muscles without tensing them first.

If you would like to learn more about PMR before trying it, the website for the U.S. Department of Veterans Affairs (VA) Whole Health Library (www.va.gov/WHOLEHEALTHLIBRARY) has information on PMR and other body-based approaches. Go to the *Tools* page and scroll down to the section titled *Power of the mind*.

3. Use your breath. You would think we would be experts at breathing since we have been doing it constantly for as long as we have been alive. However, chronic stress and its accompanying physiological response can cause us to breathe in less than optimal ways.

If you have never tried a breathing exercise, you might want to start by reading about and practicing *diaphragmatic breathing*. Again, the VA Whole Health Library site has helpful information to get you started, and there are many tutorials online.

Other forms of breathing exercises that you might want to try include the following:

- Square breathing—also called box breathing, or 4 x 4 breathing—refers to the practice of inhaling for a count of 4, holding your breath for a count of 4, exhaling for a count of 4, and pausing for a count of 4 before starting the cycle again.
- 4–7–8 breathing involves inhaling for a count of 4, holding your breath for a count of 7, and exhaling for a count of 8, repeating it for four cycles.

Some online tutorials include, in addition to audio guides, visual aids that you might prefer to counting.

4. Try a grounding exercise. "Grounding" techniques are meant to bring you back to the present moment, re-focus your attention, and calm yourself by bringing your awareness back to the body when you get swept up by the stress of the moment.

The 5–4–3–2–1 exercise is a helpful one that focuses on your senses. You can engage in it in many different moments and settings, and you can perform it privately even when in public. It involves identifying:

- 5 things you can see,
- 4 things you can hear,
- 3 things you can feel,
- 2 things you can smell, and
- 1 thing you can taste.

It is a simple exercise, but you can find sample guides online, some with minor variations.

5. Try a guided mindfulness meditation. You will find an almost endless supply of exercises labeled "mindfulness meditation" online, many with indications like "for anxiety," "for depression," "for sleep," "for rumination," and even "for productivity." Many are quite similar to the body scan, breathing exercises, or grounding exercises mentioned above. Some are as brief as a couple of minutes, and some as long as 30.

Ultimately, however, mindfulness is a practice, and if you feel it might be something you decide to pursue on a more regular basis, below are the experts I recommend. I am listing their websites, which contain valuable resources, but they also have some of the best published books on mindfulness and other forms of meditation, they are all on social media (where they post videos and recommendations), some have podcasts, and some have apps:

- Sharon Salzberg (www.sharonsalzberg.com)
- Tara Brach (www.tarabrach.com)
- Jon Kabat-Zinn (www.jonkabat-zinn.com)
- Jack Kornfield (www.jackkornfield.com)

I have also found the books *Good Morning, I Love You* by Shauna Shapiro and *The Mindful Way Through Depression: Freeing Yourself from Chronic Unhappiness*, by Mark Williams and others (not just for people experiencing a depressive episode), particularly informative and helpful. The *Healthy Minds Program* (www.hminnovations.org), also available as an app, is a well-being method by neuroscientist Richard Davidson that incorporates mindfulness-based approaches and other skills training, through practical exercises and science-based information. Finally, the VA also has a free app, *Mindfulness Coach*, with information, guided meditations, and options to track your practice and progress.

6. *Use the mind's eye.* Imagery has long been used by athletes, musicians, and even surgeons to enhance performance. Vividly imagining yourself successfully performing a challenging act, like giving a presentation, making it calmly through a medical procedure, or completing a challenging workout, can help your performance, increase your confidence, and enhance positive emotions.[11] It can be helpful to tie your imagery to a cue—for example, if you are anxious about a presentation you are giving at the end of the day, every time you catch yourself fidgeting nervously, you practice the imagery of yourself successfully presenting.

7. *Move.* We mentioned the importance of movement when talking about pauses, and we will discuss it again in Chapter 20 when going over strategies to increase alertness. Even simple stretches and a short, gentle walk can serve as cues to mind the body. You can also try a mindful walking meditation. There are different ways to do this, from focusing on sensory input (for an example, see the information on mindful walking at www.positivep sychology.com) to focusing on the actual movement of your gait (for an example, see the guide at the Greater Good in Action website, at https://ggia.berkeley.edu).

8. *Talk about it.* Normalize minding the body throughout your day. While many of these practices are performed individually, you can point out to

others, for example, when you have all been sitting for too long. Stand up and pace a little when a meeting has been going on for a while. Invite others to join you on a short walking break. You can even propose a walking meeting, if appropriate. Model for your children what it looks like to be mindful of your body, and ways to re-synchronize with it.

References

1. Gibson, A. (2018). Tincture. In: *Lord of the butterflies*. Button Publishing.
2. Britton, W.B. (2019). Can mindfulness be too much of a good thing? The value of a middle way. *Current Opinion in Psychology*, 28, 159–165.
3. Quoidbach, J., Mikolajczak, M., & Gross, J.J. (2015). Positive interventions: An emotion regulation perspective. *Psychological Bulletin*, 141(3), 655–693.
4. Diaz-Silveira, C., Santed-Germán, M.-A., Burgos-Julián, F.A., Ruiz-Íñiguez, R., & Alcover, C.-M. (2023). Differential efficacy of physical exercise and mindfulness during lunch breaks as internal work recovery strategies: A daily study. *European Journal of Work and Organizational Psychology*, 32(4), 549–561.
5. Schumer, M.C., Lindsay, E.K., & Creswell, J.D. (2018). Brief mindfulness training for negative affectivity: A systematic review and meta-analysis. *Journal of Consulting and Clinical Psychology*, 86(7), 569–583.
6. Johnson, S., Gur, R.M., David, Z., & Currier, E. (2015). One-session mindfulness meditation: A randomized controlled study of effects on cognition and mood. *Mindfulness*, 6, 88–98.
7. Polizzi, C.P., Baltman, J., & Lynn, S.J. (2019). Brief meditation interventions: Mindfulness, implementation instructions, and lovingkindness. *Psychology of Consciousness: Theory, Research, and Practice*, 9(4), 366–378.
8. Poissant, H., Mendrek, A., Talbot, N., Khoury, B., & Nolan, J. (2019). Behavioral and cognitive impacts of mindfulness-based interventions on adults with attention-deficit hyperactivity disorder: A systematic review. *Behavioural Neurology*, 2019, Article 5682050.
9. Whitfield, T., Barnhofer, T., Acabchuk, R., Cohen, A., Lee, M., Schlosser, M., Arenaza-Urquijo, E.M., Böttcher, A., Britton, W., Coll-Padros, N., Collette, F., Chételat, G., Dautricourt, S., Demnitz-King, H., Dumais, T., Klimecki, O., Mei-berth, D., Moulinet, I., Müler, T. ... Marchant, N.L. (2022). The effect of mindfulness-based programs on cognitive function in adults: A systematic review and meta-analysis. *Neuropsychology Review*, 32, 677–702.
10. Jeffay, E., Ponsford, J., Harnett, A., Janzen, S., Patsakos, E., Douglas, J., Kennedy, M., Kua, A., Teasell, R., Welch-West, P., Bayley, M., & Green, R. (2023). INCOG 2.0 guidelines for cognitive rehabilitation following traumatic brain injury, part III: Executive functions. *Journal of Head Trauma and Rehabilitation*, 38(1), 52–64.
11. Rhodes, J., & May, J. (2022). Applied imagery for motivation: A person-centred model. *International Journal of Sport and Exercise Psychology*, 20(6), 1556–1575.

20 Sparking Alertness

I was surprised to learn that the origins of the word "lethargy" go back to the ancient Greek words *léthē* meaning "forgetfulness" and *argós* meaning "idle," an early recognition of the fundamental relationship between a sedentary physical state and cognitive glitching, between our physical and mental settings. It is hardly surprising that difficulties maintaining alertness are such a common complaint, given the prevalence of powerful forces that deplete our energy, including sedentary days, chronic sleep deprivation and sleep disorders, psychological conditions like depression, medical conditions and medication side effects, and substance use. While we have a limited ability to quickly and effectively replenish our mental energy reserves, even a small boost to our alertness can significantly reduce disruptive cognitive glitches.

The Natural Rhythms of Alertness

Many factors affect our alertness during the day, including our sleep the night(s) before, our food intake, our posture, the environmental temperature, our caffeine consumption, our level of physical activity, and lighting conditions. All of these factors can also affect our baseline attentional levels.[1]

Alertness is tightly linked to the sleep–wake cycle. Overall, the sleep–wake cycle is regulated by two mechanisms.[2] *Homeostatic sleep pressure* increases with time spent awake: The longer we have been awake, the sleepier we feel. Under normal circumstances, and assuming we wake up in the early morning, our homeostatic need for sleep is low throughout the morning but increases during the day and is highest in the evenings, after we have been awake for a long time. The second mechanism is the *circadian "pacemaker,"* which influences wakefulness and sleep propensity in a cycle lasting approximately 24 hours. Our circadian rhythm starts increasing its alerting signals a few hours before we wake up. These alerting signals continue to rise, and they peak by the early afternoon in most people.[3] Circadian alerting signals then start decreasing and are lowest in the late evening.

At any given time, our level of sleepiness is determined by the interaction between these two forces. In the late morning, for example, we have only

DOI: 10.4324/9781003409311-24

been awake for a few hours, so we have very little homeostatic need for sleep. Similarly, the circadian rhythm's alerting signals are still on the rise. As a result, we feel pretty awake. In the early evenings, we have been awake for a long time, so there is high homeostatic sleep pressure. However, the circadian rhythm's influence on alertness is not low enough yet to make us fall asleep. It is later in the evening, when we experience a high homeostatic need for sleep and the circadian rhythm is powering down alertness, that we are much more likely to fall asleep. Again, this is all under normal circumstances and assuming no significant sleep deprivation, night shift work, or other disruptive forces.

Another powerful factor influencing alertness is our *chronotype*, or our natural preference for the timing of sleep and activity.[4] As is commonly known, some of us, morning types or "larks," tend to wake up early, feel mentally and physically at our best in the morning, and have difficulty remaining awake past our bedtime.[2] Some of us, evening types or "owls," tend to wake up later, perform and feel best in the late afternoon and evening, and can have extreme difficulty getting up early in the morning. This preference is rooted in biology: Our chronotype actually correlates with physiological and behavioral measures of circadian arousal.

Chronotype changes throughout life.[4] Children tend to be morning types, but in puberty there is a shift toward eveningness. By college age, about 30 to 40 percent of us continue to be evening types, while 50 to 60 percent shift to a neutral preference, and only a very small minority show morningness tendencies. The shift toward morningness continues through middle age, and by our 60s, about three-fourths of us are morning types, while the rest tend to be neutral types, without strong morningness or eveningness tendencies. Very few of us remain strong evening types in older age.[2]

The Natural Rhythms of Our Brains

In general, cognitive abilities decrease the longer we have been awake.[5] As sleepiness increases and alertness decreases, cognitive performance also decreases. Most of the literature on this, however, has evaluated basic aspects of attention: For the most part (when we are not sleep deprived), attentional abilities increase from 7 a.m. to 11 a.m., remain relatively high until around 3 p.m., then begin to steadily decline.[6] This is true based on self-reported attention, but also performance on tests and physiological measures including EEG. Some factors can make a difference—for example, tasks that are highly motivating and brief allow us to "rally" and increase our alertness and attention levels.

As you might be anticipating, however, the timing of our cognitive performance peak does vary depending on our chronotype.[2] Overall, we perform the best cognitively when we are tested at our preferred time of day based on our chronotype: Morning types perform better early in the day, and evening types perform better later in the day.[4] This is referred to as the *synchrony effect.*

The synchrony effect is surprisingly powerful. Synchrony (better performance at our preferred time of day) results in improvements in attention, resistance to distraction, learning, recall, working memory, prospective memory, and other executive functions. The effects seem to be more significant when effortful or strategic processing is required, greater working memory is needed, there is greater need to inhibit distractions, and we need to inhibit automatic responses to familiar stimuli—in other words, whenever significant executive control is needed to override habit and act deliberately and thoughtfully. Our reasoning and decision-making is also better at our preferred time of day: We are more likely to analyze and respond to complex arguments based on detailed information, while during our "off-peak" times, we are more easily persuaded by celebrity endorsements, engaging images, flattery, or attractive speakers. During our off-peak times, we are also more likely to make flawed decisions based on logical shortcuts and stereotypes, make mistakes, and even behave unethically.

Synchrony is less important for tasks that rely on well-learned, automatic responses. Interestingly, for some kinds of creative problems that require "thinking outside the box," we can perform better at off-peak times, probably because we are less likely to be constrained by logic.

Synchrony effects are more pronounced in people with cognitive impairment from neurological conditions, suggesting we are also more vulnerable to synchrony effects when we are cognitively depleted. In other words, when we are already depleted, we might perform even worse during our off-peak times. Synchrony effects are also larger in older adults, who show exaggerated deficits in performance when tested at off-peak times.[2], [4] Moreover, in older age, even neutral types tend to show better cognitive performance in the morning and mid-day.

The improvement in our cognitive performance at our preferred times of day has a physiological basis: During our preferred, "peak" times, we have higher levels of cortical excitability and facilitation.[7] This means that the neurochemical and electrophysiological processes that make it more likely that neurons will fire, that facilitate learning and the formation of new memories, and that underlie neuroplasticity, are stronger during our preferred times. In other words, during our preferred times, our brains are physiologically primed for learning and performance.

What You Can Try Today

Before we dive in, I want to be very clear about something: Sleepiness is one major factor affecting alertness, but it is not the only one. The strategies below will be of limited, if any, help if you are experiencing decreased alertness due to acute sleep deprivation. When you are sleep deprived, the only solution is sleep. There is no effective antidote for sleep deprivation, other than sleep. These strategies cannot, for example, help you drive safely or prevent you from falling asleep while babysitting if you slept four hours last night. These strategies are meant to help you manage the natural ebbs and flows of alertness, which might be accentuated by being cognitively or physically depleted or overwhelmed.

1. Drink a cup of coffee. (Or caffeinated tea.) Yes, caffeine does increase alertness; it amplifies cortical activity in brain regions important for sleep, cognition, and mood.[7] Consuming a standard cup of coffee significantly improves memory in young and older adults during their non-preferred times (early mornings in young college students and afternoons in older adults), potentially by as much as 30 percent.[8], [9] Caffeine also improves attention and reaction time, but it has less consistent effects on complex executive functions like judgment and decision-making.[10]

A couple of important caveats. First, caffeine does not give our memory a boost during our peak times; it is only effective in improving cognitive functioning when our physiological arousal levels are low. Second, the studies have been performed in individuals who were regular caffeine drinkers. If you are not a caffeine drinker, you need to be mindful of the fact that you might initially experience sone negative side effects like shakiness, anxiety, and, paradoxically, difficulty concentrating if you drink caffeine. Finally, remember from Chapter 7 that levels of caffeine peak about 30 minutes after we drink it, and in most of us caffeine has a half-life of 5–7 hours, so time your caffeine consumption wisely and avoid consuming it after 3 p.m. to avoid interference with your sleep.[3]

2. Take a short nap. Naps are an effective way to increase alertness by reducing homeostatic sleep pressure.[5] As we have mentioned before, as a result, naps can improve executive functioning, learning and memory formation, and emotional processing. If you have a cognitively demanding day, a nap in between cognitive activities can improve your memory.[11]

If you can, take a short nap of between 10 and 20 minutes.[12] Use an alarm to make sure you do not sleep any longer. (Despite my own sleep problems, I was skeptical about "cat naps" for years. I assumed I would not feel refreshed, and I feared that I would perhaps feel worse. As predicted by the scientific literature, I was wrong, and I am now a convert and a big fan of cat naps.)

Remember that longer naps, when we actually go into deep non-REM sleep, result in *sleep inertia*, a period of disorientation and cognitive sluggishness immediately after awakening from deep sleep.[5] During sleep inertia, we actually experience decreases in attention and other cognitive functions.[1] So keep your naps short. Also, avoid napping after 3 p.m., to avoid interference with your night-time sleep.

3. Move. If you feel your cognitive processing becoming sluggish, change your body position, or, ideally, stand up, walk, and stretch. Maybe even play an energizing song and move to the rhythm for a bit.

4. Seek some natural light. As we reviewed in Chapter 7, light has a powerful effect on alertness, which is why exposure to bright light is a common culprit of our difficulties falling asleep. Photoreceptors in your eyes transmit light information to areas of the brain important for the regulation of alertness, including the suprachiasmatic nucleus (which has been called the "master pacemaker" of the brain) and the hypothalamus.[13] Exposure to light increases feelings of alertness and improves attention. So turn up the light in your environment if possible, or, even better, go outside or stand in a spot with plenty of bright natural light.

5. *Eat the right snack.* Hypoglycemia, or low blood sugar levels, can reduce working memory and attention span.[14] Unsurprisingly, then, elevations in glucose from eating result in relatively fast, short-term improvements in working memory. Moreover, combining glucose and caffeine has been found to improve working memory better than either alone. However, foods that are high in sugar, like energy drinks and sugary snacks, can cause us to "crash," worsening our lethargy and cognitive lapses.[15] Eat small meals throughout your day to avoid acute fluctuations in your blood sugar levels, and if you need a boost, have some protein and some healthy sugar (e.g., from fruit), but avoid highly processed sugary snacks that cause steep increases in blood sugar followed by crashes.

6. *Talk about it.* Normalize fluctuations in alertness and the need to intentionally address its ebbs and flows. Stop pretending your alertness does not wane during a long afternoon meeting. Stand up while continuing to participate, suggest going for a walk outside prior to the meeting, grab caffeinated drinks and healthy snacks for everyone, and make sure the room is brightly lit.

References

1. Hudson, A.N., Van Dongen, H.P.A., & Honn, K.A. (2020). Sleep deprivation, vigilant attention, and brain function: A review. *Neuropsychopharmacology Reviews*, 45, 21–30.
2. Schmidt, C., Collette, F., Cajochen, C., & Peigneux, P. (2007). A time to think: Circadian rhythms in human cognition. *Cognitive Neuropsychology*, 24(7), 755–789.
3. Walker, M. (2017). *Why we sleep: Unlocking the power of sleep and dreams.* Scribner.
4. May, C.P., Hasher, L., & Healey, K. (2023). For whom (and when) the time bell tolls: Chronotypes and the synchrony effect. *Perspectives on Psychological Science*, 18(6), 1520–1536.
5. Mantua, J., & Spencer, R.M.C. (2017). Exploring the nap paradox: Are mid-day sleep bouts a friend or foe? *Sleep Medicine*, 37, 88–97.
6. Kraemer, S., Danker-Hopfe, H., Dorn, H., Schmidt, A., Ehlert, I., & Herrmann, W. M. (2000). Time-of-day variations of indicators of attention: Performance, physiologic parameters, and self-assessment of sleepiness. *Biological Psychiatry*, 48(11), 1069–1080.
7. Salehinejad, M.A., Wischnewski, M., Ghanavati, E., Mosayebi-Samani, M., Kuo, M.F., & Nitsche, M.A. (2021). Cognitive functions and underlying parameters of human brain physiology are associated with chronotype. *Nature Communications*, 12(1), Article 4672.

8. Sherman, S.M., Buckley, T.P., Baena, E., & Ryan, L. (2016). Caffeine enhances memory performance in young adults during their non-optimal time of day. *Frontiers in Psychology*, 7, Article 1764.

9. Ryan, L., Hatfield, C., & Hofstetter, M. (2002). Caffeine reduces time-of-day effects on memory performance in older adults. *Psychological Science*, 13(1), 68–71.

10. McLellan, T.M., Caldwell, J.A., & Liebeman, H.R. (2016). A review of caffeine's effects on cognitive, physical, and occupational performance. *Neuroscience and Behavioral Reviews*, 71, 294–312.

11. Leong, R.L.F., & Chee, M.W.L. (2023). Understanding the need for sleep to improve cognition. *Annual Review of Psychology*, 74, 27–57.

12. Summer, J., & Singh, A. (2024, March 11). Napping: Benefits and tips. Sleep Foundation. www.sleepfoundation.org/napping.

13. Lok, R., Smolders, K.C.H.J., Beersma, D.G.M., & de Kort, Y.A.W. (2018). Light, alertness, and alerting effects of white light: A literature overview. *Journal of Biological Rhythms*, 33(6), 589–601.

14. Blasiman, R.N., & Was, C.A. (2018). Why is working memory performance unstable? A review of 21 factors. *Europe's Journal of Psychology*, 14(1), 188–231.

15. Sternberg, E.M. (2023). *Well at work. Creating wellbeing in any workspace*. Little, Brown Spark.

21 Reclaiming Your Time

Every strategy we have discussed has one thing in common: It takes time. The only way to build a brain-friendly life is to invest the time to implement the changes that help replenish cognitive resources and decrease cognitive demands. And yet, when we are depleted and overwhelmed, time seems to be in shorter supply than ever. Taking the time to organize our days to create an environment in which our depleted brains can actually function (and ideally be less depleted) can lead to a less stressful and healthier life by decreasing glitches, frustration, and chaos.

A Depleted Brain Is a Chaotic Brain

As we have seen repeatedly throughout the book, a depleted brain struggles particularly with the cognitive functions that allow us to organize our behavior in flexible and adaptive ways so we can meet our goals—our executive functions. Unsurprisingly, when we are cognitively drained, our internal and external worlds can feel chaotic. The reasons include all of the following:[1]

- We are more distractible. We have difficulty staying on task, because our attention is easily pulled by whatever is going on in our environment. We get sidetracked by other activities, leave a trail of unfinished chores, and lose track of what we were doing. We forget things because we are not paying attention, and we do not follow through with things we intended to do.
- We have difficulty initiating tasks. We cannot "get going." We *procrastinate*, meaning we delay tasks voluntarily but unnecessarily.[2] We intend to do something, we know it is important, and yet we simply do not get it done. Procrastination is not a problem with time management or lack of motivation: It is a counterproductive attempt at emotion regulation. We procrastinate tasks that we find unpleasant, boring, or anxiety-provoking. We avoid the task to avoid the emotion. And just as with other unhealthy coping approaches, we are more likely to procrastinate when depleted, because it brings immediate "relief" even

DOI: 10.4324/9781003409311-25

though it causes a problem for the future and prevents us from achieving our goals.

- We are cognitively inflexible. We act out of habit, and we cannot see different ways to do things. We have a difficult time prioritizing, and instead do whatever is right in front of us. We have difficulty shifting priorities when something new comes up, adapting when plans change, or coming up with a new plan if something unexpected happens. We perseverate, we get "stuck" in one problem or a single solution to a problem, and we have difficulty thinking of creative solutions.
- Our ability to plan is weakened. When we are depleted, we become more concrete, so it is more difficult to look ahead and consider goals far in the future. We look at the list of things we need to accomplish, we do not know where to start, and we do not have the capacity to figure it out. If we feel rushed, we do not plan because we feel we do not have the time, so instead we stumble through the day tackling whatever activity or chore seems most urgent next.
- Our time management is poor. Time estimation is an executive function, and when our executive resources are depleted, we are less accurate at estimating how long a task will take. Most commonly, we under-estimate, so we do not allocate enough time to complete a task, we try to squeeze one more task before our next activity, and we end up run-ning late, leaving things unfinished, and increasing chaos and over-whelm. We can also perceive deadlines as further than they actually are, so we do not start working on things until it is too late.
- We are less self-aware. We have difficulty seeing clearly how we are doing, estimating whether we are still on track to meet our goals, and catching our mistakes. We are less "in tune," lose track of time, and get caught up in what we are doing, not noticing that it is time to do something else.

The main way to compensate for these weakened executive functions involves organizing our time so that we can, deliberately and con-scientiously, create external supports for our depleted internal resources.

What You Can Try Today

1. Set reasonable goals for today. Setting goals is an effective technique to improve functioning in daily life. In fact, training in goal setting and goal management are important elements of cognitive rehabilitation programs.[3] Like those experiencing executive dysfunction due to neurological pro-blems, when we are depleted and our executive functions are weak, we have difficulty setting meaningful and achievable goals. The purpose of set-ting goals is to decrease cognitive demands by creating a roadmap for our days, instead of having to make repeated decisions during the day about what we should be accomplishing.[4]

When setting your goals for today, remember the important and urgent matrix from Chapter 17, and prioritize those goals that are both important and need to be accomplished soon. Try to be specific and realistic: Goals like "Have a productive day" or "Be more active" are not particularly helpful because they are quite vague. Similarly, "Organize and file all the household paperwork, clean the house, and start a new strength training program" might be too ambitious. Choose goals that seem achievable for your physical and cognitive resources and the demands in your life. As I write this, for example, my goals for tomorrow are attending one workout class in the morning, preparing half of the material I need for the weekly class I teach, and finishing one of my pending neuropsychological reports. Those are my priorities for the entire day, because I know there will be a dozen other tasks that will pop up once I get to work.

2. Plan. We often think we do not have time to make a neat, color-coded plan, set reminders, and pause to review our plan in the middle of the day. In fact, planning often gets neglected when we are overwhelmed, rushing, or sleep deprived. That is unfortunately a self-defeating decision. Planning reduces time pressure and stress, resulting in a decreased chance of errors, so it is time well spent.[5] Before starting your day, plan the day by creating a schedule and to-do list (see below) based on your goals. If you are going grocery shopping, create your shopping list. If you are going to an appointment in an unfamiliar part of town, look up directions and save them (or print them) in advance. If you are taking your car to the shop, gather your paperwork and write down your questions. Plan your meals for the day.

3. Make a to-do list. Based on your goals, create a to-do list that includes not just a list of tasks you need to accomplish, but how long you estimate each task is going to take. When making your to-do list, make sure to break up tasks. A principle for cognitive compensation and rehabilitation is that when we cannot change our cognitive functioning, we need to change the task.[7] By breaking up the task into components, we make it less complex; this helps with our limited cognitive resources and motivation, which can flag halfway through longer tasks.[8]

For example, for my "prepare half the material for class" goal, the actual items on my to-do list will include: grade last week's assignments, put together the weekly quiz, put together the slides for the first hour of class, look up updated educational resources on the brain's blood supply system, and put together materials for an in-class activity, each with the estimated time I expect the task to take. Similarly, instead of writing "food prep" on your to-do-list, you might include: make a meal plan, make an ingredient list, go to the grocery store, cook meal #1, cook meal #2, and divide portions and store in labeled containers. (Depending on your level of depletion, you might need to break the tasks down into even simpler components.) Breaking up tasks like this also helps with time estimation, because it is easier to estimate how long it takes us to go to the grocery store than how long it will take to "food prep."

It is okay if your to-do list includes things you know you will not get to today. It is helpful to keep a running list of activities that will take five minutes (call to schedule a dentist appointment), 15 minutes (look up and review a hospital bill), 30 minutes (return an item to a store), an hour (complete the online training required for work), two hours (clean the pantry), and even a day (clean the garage) to complete. During the day, when you find yourself with five, ten, or 30 minutes to spare before your next activity, you can turn to this time-based to-do list and pick things to do based on how much time you have available.

Notice that this strategy does introduce some shifting, so only tackle the task if you will have enough time to pause and shift before your next activity. Similarly, under some circumstances, using that extra time for a restorative or executive pause—to review what you have accomplished so far, to engage in a pleasant or relaxing activity, to do a quick mindfulness or relaxation exercise, to walk or stretch—might be a better use of that time, depending on what you need most in that moment.

4. Make a schedule. We have mentioned that self-monitoring is an executive function, so when we are depleted, our ability to accurately assess how we are doing tends to suffer. Because of this, it might be difficult for us to accurately assess how we are doing in terms of achieving our goals. Do not just write down what you need to accomplish, but make a schedule with start and stop times for each task so that, as the day goes by, you can track how you are doing—whether you are staying on track or you are falling behind and need to revise you goals and plans.

This level of detail might sound like overdoing it, but when we are depleted and overwhelmed and we do not write tasks down on a schedule with external or self-imposed deadlines, we miss things, delay activities, and pay a price: We are too busy to make a medical appointment for that pain we have been feeling, and then we end up in the ER. We keep postponing making a minor car repair, and end up making the problem worse and spending a large sum of money on a larger repair. We do not make time to go over our finances, so we miss due dates and pay interest and late fees. Scheduling and monitoring our schedules is one of the most effective things we can implement in our daily lives to reduce glitches and improve our overall functioning.

When making a schedule:

- Schedule a few easy tasks first. Early in a busy or overwhelming day, tackle some short and relatively low-effort tasks. Staring at a long to-do list can feel daunting, worsen your mood, and actually make you feel paralyzed. Start your day with a couple of easily achievable tasks and cross them off, to give your mood a boost and fuel your self-efficacy and motivation to get more things done.
- Capitalize on your normal rhythms. Remember chronotypes and synchrony. When do you feel more alert, energized, and productive? In the

early mornings, late mornings, afternoons, or evenings? Keep this in mind when creating your schedule. If you are a morning person, schedule more breaks in the afternoon, and schedule your more cognitively demanding tasks during your peak times. For example, if you are very much a morning person, and you have to finish your child's play costume today, you want to schedule working on the costume early in the morning if you are not particularly skilled at sewing, but if you are an expert seamstress and this is a relatively easy, low-effort task for you, you should schedule it for later in the day, when you might be more cognitively depleted and unable to tackle more cognitively demanding tasks.

- Plan for breaks. Remember the importance of breaks from Chapter 12. Include protected time for breaks in your schedule. Leave some time in between activities. Give yourself ample time, especially if you think an activity will be stressful and depleting.
- Include "executive" breaks to check your work and to check your progress towards your goals. Remember that when depleted or stressed, we are more likely to make mistakes due to attentional lapses, and we are also more likely to not notice our mistakes. If you have a busy day or are doing several important things, schedule breaks devoted to checking—check that you wrote all the appointments you made on your calendar, check that all the payments are being sent to the right place, check you cc'd everyone on the email you sent.
- Set reminders as needed—for example, ten minutes before a meeting starts, so you have time to shift and prepare, and five minutes before a meeting has to end, so you can start wrapping up without rushing.

5. Protect your routines. A recurrent theme throughout the book has been that activities that are automatic and habitual are cognitively cheap, while activities that are novel are cognitively expensive and require more executive resources. Moreover, when we are depleted, we become cognitively inflexible, and when we are stressed, we resort to habit. Because of this, when we are depleted, we can significantly reduce glitching by protecting our routines. For example, your mornings might be very busy, with quite a bit to do and a lot of shifting, but you might have a well-rehearsed routine to get you through those couple of hours between waking up and dropping your children off at school. Do not disrupt that routine. If somebody texts you, "Can I call you super quick just for a minute?" reply, "I can talk at 8:30," or whenever you will be done with your morning routine.

6. *Mark the end of your day.* Before calling it a day, take some time to go over everything you accomplished. Review what you did and rehearse important information so you can remember it later. Clear clutter. Make sure you know where your keys, wallet, and other important items are. Gather paperwork, receipts, and notes from medical appointments. You do not need to organize and file everything right away, but keep a bin or tray

where you put important items until you have a chance to file them away. Update and organize your notes.

7. Talk about it. Normalize relying on schedules and plans, and protecting routines. At work, encourage coworkers to keep a centralized, shared schedule of important events. At home, keep a family calendar and hold "family planning time"—for example, on Sunday evenings—when you review the calendar together and anticipate and plan for busy days. Acknowledge the limits to your availability, saying, for example, "Yes, I'll be happy to join, but I can only stay for 30 minutes."

References

1. Dawson, P., & Guare, R. (2016). *The smart but scattered guide to success.* Guilford Press.
2. Mills, K. (Host). (2022, October 12). Why we procrastinate and what to do about it, with Fuschia Sirois, Ph.D. [Audio podcast episode]. Speaking of Psychology. American Psychological Association.
3. Elbogen, E.B., Dennis, P.A., Van Voorhees, E.E., Blakey, S.M., Johnson, J.L., & Johnson, S.C. (2019). Cognitive rehabilitation with mobile technology and social support for veterans with TBI and PTSD: A randomized clinical trial. *Journal of Head Trauma Rehabilitation*, 34(1), 1–10.
4. Swann, C., Jackman, P.C., Lawrence, A., Hawkins, R.M., Goddard, S.G., Williamson, O., Schweickle, M.J., Vella, S.A., Rosenbaum, S., & Ekkekakis, P. (2023). The (over)use of SMART goals for physical activity promotion: A narrative review and critique. *Health Psychology Review*, 17(2), 211–226.
5. Fasotti, L., Kovacs, F., Eling, P.A.T.M., & Brouwer, W.H. (2000). Time pressure management as a compensatory strategy training after closed head injury. *Neuropsychological Rehabilitation*, 10(1), 47–65.
6. Cantor, J., Ashman, T., Dams-O'Connor, K., Dijkers, M.P., Godon, W., Spielman, L., Tsaousides, T., Allen, H., Nguyen, M., & Oswald, J. (2014). Evaluation of the Short-Term Executive Plus intervention for executive dysfunction after traumatic brain injury: A randomized controlled trial with minimization. *Archives of Physical Medicine and Rehabilitation*, 95, 1–9.
7. Ponsford, J., Velikonja, D., Hanzen, S., Harnett, A., McIntyre, A., Wiseman-Hakes, C., Togher, L., Teasell, R., Kua, A., Patsakos, E., Welch-West, P., & Bayley, M.T. (2023). INCOG 2.0 guidelines for cognitive rehabilitation following traumatic brain injury, part II: Attention and information processing speed. *Journal of Head Trauma Rehabilitation*, 38(1), 38–51.
8. Alter, A. (2023). *Anatomy of a breakthrough: How to get unstuck when it matters most.* Simon & Schuster.

22 Parting Thoughts: Looking Upstream

One of my favorite cartoons ever shows a woman being burned at the stake, with another woman standing nearby asking her, "Have you considered taking up yoga?"[1]

My main concern as I thought about writing this book was the same concern I always have when I discuss my recommendations with patients: I know these strategies work, and I know this person would benefit from them, but is my advice a drop in the bucket of this person's life?

There is a quote, often attributed to Desmond Tutu (but to the best of my knowledge of unconfirmed origin), that goes, "There comes a point where we need to stop just pulling people out of the river. We need to go upstream and find out why they're falling in." I am very aware that the reasons many people are living brain-unfriendly lives go beyond the personal, and involve forces at the family, community, social, and economic levels. In that context, it can feel a little as though I am throwing floating devices at the people in the river, cheering on them to swim harder, instead of going upstream and stopping them from falling in.

While that is the nature of my job, to help one person at a time as they come through my door, I also want to point out that what I see over and over is people who turn their concern inward and assume something is wrong with *them*, when sometimes the only thing that is wrong is that they live a life full of forces that push them into the river. They work in systems designed for productivity with little regard for their well-being and sometimes their humanity. They carry heartbreak and trauma, sometimes since childhood, but have faced lifelong barriers—financial, familial, cultural—to access mental care. They are exhausted, physically and mentally, but they feel they cannot stop to rest or take care of themselves, because they carry the well-being of others on their shoulders. Addressing those forces is above my pay grade, but I do think there are things we can do, besides taking care of our health and implementing strategies to decrease our cognitive lapses as described in this book.

First, we need to become aware that many of the things we are getting wrong as institutions, communities, and societies can indeed change. We need to stop acting as though unsustainable demands, disregard for physical

DOI: 10.4324/9781003409311-26

comfort and mental health, and brain-unfriendly (and thus mind-unfriendly) days are normal. We have to stop acting as if going through life chronically stressed, sleep deprived, without time or energy to exercise, prepare healthy meals, or connect with loved ones is just the way things are when you are an adult. We have to stop saying things to each other like "I haven't had time to eat" or "I worked straight through the weekend" with a straight face. If we say it with a pinch of pride, that is even more concerning.

Second, we need to talk about this, and encourage others to join us in creating brain-friendlier days. You probably noticed that every chapter in Part III ended with a recommendation to "Talk about it." We have to normalize prioritizing our brain health, which means prioritizing our health in general. Talk about it with your children, coworkers, the other parents at school, your golf buddies, your friends. Say out loud, "I'm trying not to rush," "I'm working on taking breaks," and "I need to go on a walk." Perhaps if we sustain these relatively small changes, we can spark cultural changes at least in certain spheres.

What might this look like? Borrowing somewhat liberally from Christina Maslach's work, here are a few questions related to the six factors she has found to be related to our risk for burnout:[2]

- Do you have enough psychological resources to meet the demands on your time, attention, and energy?
- Do you feel you have choice or control over what you do and how you spend your time?
- Do you feel rewarded, recognized, and appreciated for what you do?
- Do you have a community of supportive people that you are in regular contact with?
- Do you feel that you are treated with fairness and respect?
- Do you feel that you are living according to your values and that what you do is meaningful?

Even if (especially if) we spend our days within systems that do not provide us with these protective factors, perhaps we can intentionally find ways to do this for each other. We can take turns taking some of the load off one another's shoulders, celebrating each other, and creating opportunities to rest and replenish our brains. Our social nature and the ability of our brains to pool cognitive resources to tackle challenges as a tribe is one of our greatest strengths as a species. Relying on each other to protect our brains makes perfect sense.

* * *

In an interview about her research on burnout, Dr. Maslach used the metaphor of the canary in the coal mine in a way that stuck with me because it applies to our lives in general. She said, if the canary cannot survive in the coal mine, "It's not that something is wrong with the canary, that it's not tough enough."[3] The problem is the presence of toxic fumes. The canary's demise "is a signal that something needs to be fixed." She explained that, similarly, burnout is a signal of a toxic situation.

For some of us, our glitches are a signal that our internal systems are depleted. For others, they are a signal of excessively demanding external circumstances. For many of us, they are a signal that a combination of both is at play. Our glitches are never a signal, however, that we are not tough enough or that we need to try harder, give even more, or "hustle." The solution to a healthy brain's glitches involves creating a nurturing, restorative, and sustainable environment for it, within and outside our bodies. I wish your brain a long and healthy life, full of captivating intellectual interests, awe-inspiring beauty, meaningful and passionate pursuits, rich emotional experiences, an abundant treasure trove of memories, and deeply intimate relationships, all the things that the organ upstairs thrives on, and that define what it means to be human.

References

1. Lobanova, N. (2020, August 25). This is what your unsolicited advice sounds like. *The New Yorker*. www.newyorker.com/humor/daily-shouts/this-is-what-your-unsolicited-advice-sounds-like.
2. Maslach, C., & Leiter, M.P. (2016). Understanding the burnout experience: Recent research and its implications for psychiatry. *World Psychiatry*, 15, 103–111.
3. Mills, K. (Host). (2021, July 28). Why we're burned out and what to do about it, with Christina Maslach, Ph.D. [Audio podcast episode]. Speaking of Psychology. American Psychological Association.

23 Appendix
Seeking Help: The Evaluation Process

As I mentioned in the Introduction, while I hope the information and advice in this book are reassuring and helpful, they are not meant to discourage anyone concerned about their cognitive lapses from seeking professional help. Here is some information on when and how to seek an evaluation, and about how the process might go.

When To Seek an Evaluation

When are cognitive changes concerning enough to seek neurological eva-luation? The answer will depend on many factors. In the broadest sense, anyone experiencing cognitive changes that are interfering with their func-tioning in daily life or are causing them distress should discuss their con-cerns with their physician. You will make a joint decision about whether to pursue neurological evaluation based on the nature of your cognitive diffi-culties, your medical history, and other factors. If many of the conditions discussed in Part II are present, perhaps you will decide to tackle those first and see if your cognitive functioning improves. If you are younger and healthy, perhaps you will decide to try the strategies from Part III and see if your cognitive functioning improves.

Under certain circumstances, however, it is preferable to seek neurological evaluation sooner rather than later:

1. Cognitive changes in an older adult. Age is one of the most important risk factors for neurocognitive disorders like mild cognitive impairment (MCI) and dementia. Adults in their 60s or older experiencing cognitive changes that represent a decline from their baseline and/or cause problems in their daily life should be referred for evaluation.

As we reviewed in Chapter 3, two things are true. On the one hand, subjective cognitive complaints in an older adult can be the earliest signs of concerning cognitive decline. Because of this, evaluation is recommended, to document the person's cognitive status, to identify any subtle changes, and as a baseline to assess for decline over time.

On the other hand, as we mentioned in Chapter 2, it is not uncommon for people in the early stages of a neurocognitive disorder like MCI or dementia

DOI: 10.4324/9781003409311-27

to lack insight into their cognitive impairment. (In this sense, feeling concerned about our cognitive changes can be a good sign.) If you notice significant cognitive changes in an older loved one, especially if the changes are causing problems—they are missing appointments or payments, for example—and they themselves seem unconcerned, an evaluation should be pursued.

One more point: When older adults are in the early stages of a neurocognitive disorder, they might remain able to function appropriately in highly familiar, structured, or routine situations. If we see grandma every Sunday in a structured setting like church (where we are largely following a "script"), and then we go have brunch at the same restaurant, with the same people, where she orders the same thing every time, we might not notice mild cognitive changes. It is often when we see them in a novel situation that demands flexibility and executive functioning (e.g., during a vacation or large family function) that cognitive impairment becomes evident.

2. Pronounced personality or behavioral changes. Evaluation should be sought when a person displays profound changes in their mood or behavior, with or without accompanying cognitive impairment. This includes:

- a first episode of depression later in life,
- acute and severe anxiety in somebody with no prior history,
- significant apathy leading the person to abandon most activities they previously enjoyed,
- socially inappropriate behaviors (e.g., cursing loudly at the playground or telling off-color jokes to strangers),
- impulsivity or poor judgment (e.g., falling for obvious online scams),
- pronounced irritability or lashing out that is out of character,
- auditory or visual hallucinations (hearing or seeing things that are not actually there),
- delusions (e.g., becoming convinced somebody is surveilling their home, accusing their spouse of infidelity), and
- significant sleep disturbance, including acting out their dreams (e.g., punching or "running" in their sleep) or sleeping excessively during the day.

3. Cognitive changes in individuals who are at higher risk of neurocognitive disorders. This includes cognitive decline in:

- people with neurological conditions (like Parkinson's disease, epilepsy, or a previous history of stroke) who did not display cognitive symptoms before,
- those with medical conditions that increase their risk for cognitive decline and neurocognitive disorders (like heart disease, severe alcohol use disorder, or advanced, poorly controlled diabetes),

- those living with other serious or complex medical conditions (such as cancer), and
- those with a significant family history of a neurodegenerative condition like Alzheimer's disease.

4. Cognitive changes that do not respond to supports. Even if you are a younger, healthy adult, if you have noticed cognitive changes, you have addressed the conditions from Part II so that your physical and mental health is seemingly well managed, and you have applied the strategies from Part III, yet your cognitive changes do not improve or they actually worsen, it is probably time to seek evaluation by a specialist.

5. Rapid onset of neurological changes. Of course, whenever neurological symptoms develop acutely, prompt medical evaluation should be pursued. This includes episodes of confusion, disorientation (not knowing what day it is, where they are, or who familiar people are), slurred speech, weakness, balance problems, etc.

The Evaluation Process

The evaluation process can be handled in different ways. What follows is the process that I personally recommend, based on my experience, and if the person's situation allows it.

First, talk to your primary care physician. They can do medical tests, like a blood test panel, to check for conditions like metabolic abnormalities, vitamin deficiencies, and renal and liver dysfunction, to identify or rule out conditions that can be causing or exacerbating cognitive symptoms. They can also perform a brief cognitive screening. Based on this information, the nature of your cognitive complaints, and your medical history, you can make a decision together about whether to pursue evaluation by a specialist. If you decide you should, you can then be referred to a memory disorders clinic, most likely to a behavioral neurologist with expertise in memory disorders and other cognitive changes.

Along with the referral, it is important for your primary care physician to provide the neurologist with documentation of your medical history—for example, the conditions you are being treated for, the results of any recent tests performed, and previous brain imaging studies. In preparation for your neurology appointment, it can be helpful for you to put together some other details about your medical history, including, for example, when you started noticing your problems, whether they seem to be progressing or fluctuating from day to day, your history of major injuries, illnesses, and surgeries, and a list of all medications and supplements you are taking. It is also helpful if someone close to you who has noticed the cognitive or behavioral changes joins you for the appointment.

At the neurological appointment, the neurologist will perform an interview, neurological exam, and maybe additional cognitive screening. Based on their initial impressions, they might order additional tests, most likely a brain MRI to examine the structural health of your brain, including, for example, patterns of atrophy or the presence of cerebrovascular changes. Under certain circumstances, depending on your history and symptoms, other studies, like a PET scan, can be ordered, to also examine the brain's metabolic activity. The neurologist might also refer you for neuropsychological evaluation.

It is important for you to know that the diagnostic process for neurocognitive disorders is rapidly evolving. As I write this, much progress is being made in the study of biomarkers for the clinical diagnosis of conditions like Alzheimer's disease and other dementias. Such diagnostic tools might be available in the not-so-distant future and change how we evaluate people for these disorders.

The neuropsychological evaluation typically lasts several hours, and it can happen all in one day or over a couple of sessions. It involves an in-depth interview, ideally including someone who is familiar with your baseline level of functioning (what your cognitive strengths and weaknesses were before any changes) and who has also had opportunity to observe first-hand any changes in your cognition, emotions, and behavior. You will then complete standardized tests that assess multiple cognitive domains, including processing speed, attention, memory, executive functions, language, visuospatial functions, and motor functions. Emotional and behavioral functioning—like depression and anxiety symptoms—are also assessed. The specific tests administered will vary depending on your age, your specific concerns, and how you perform on tests as the evaluation progresses.

The neuropsychological evaluation allows us to do a few things. First, we can tease apart problems in different functional domains—for example, examine whether your memory problems are due to a primary memory impairment or due to slow processing speed, deficits in attention, executive function problems, or depression.

Second, by comparing your performance to that of other individuals of similar age and background, we can detect the presence of *impairment*; by comparing your performance to your estimated baseline, we can detect the presence of *decline*. As we mentioned in Chapter 3, a person can experience quite significant decline while remaining technically within normal limits, meaning their performance has not dropped enough to be in the impaired range for someone their age. Both aspects of your performance—how you are doing compared to others and how you are doing compared to your own baseline—must be considered when making conclusions about your current cognitive status and possible diagnoses.

Finally, the neuropsychological evaluation produces a profile of cognitive strengths and weaknesses. By comparing this profile with the typical cognitive profile seen in different conditions, we can then offer diagnostic impressions—for example, whether a person's profile is most consistent with that seen in Alzheimer's disease, in people experiencing the long-term consequences of a traumatic brain injury, in people with cerebrovascular changes, or in depression. This profile will also allow us to provide recommendations for treatments and other interventions and supports that might be most beneficial, like the ones presented in Part III of this book and others.

The report of the neuropsychological evaluation, which you can request a copy of, describes the results of your cognitive testing, integrates them with your history and other test results, and explains the diagnostic impressions and recommendations. The neuropsychologist can also conduct a *feedback session* with you (and a loved one, if you want them to join in), during which they review your results, explain their conclusions and recommendations, and answer your questions.

It is important to keep in mind that the neuropsychological evaluation is one piece of the diagnostic puzzle: Your referring neurologist will integrate the results of your medical history and tests, your imaging studies, the neuropsychological evaluation findings, and their own neurological evaluation, to provide you with a diagnosis and a treatment plan, if one is indicated.

Understanding a Neurocognitive Disorder Diagnosis and Figuring Out Next Steps

While I hope this is not the case for you or your loved one, given the prevalence of neurocognitive disorders, it is important to understand what a diagnosis actually means. As we reviewed in Chapter 2, the term *neurocognitive disorder* refers to disorders characterized by cognitive symptoms that affect how the person functions in daily life. There are two components to the diagnosis:

First, the actual diagnosis of a neurocognitive disorder is based on (a) how severe the cognitive impairments are, as documented on the neuropsychological evaluation, and (b) to what extent the cognitive impairments affect the person's ability to perform their activities of daily living (ADLs). A distinction is sometimes made between more complex ADLs, sometimes referred to as *instrumental* activities of daily living (IADLs) and more basic ADLs. IADLs include activities like managing finances (keeping track of expenses, paying bills on time, organizing financial paperwork for tax purposes), refilling prescriptions and taking medications as prescribed, keeping track of medical appointments, and using technology, like a smart phone or a computer. Basic ADLs include activities like those needed for hygiene, grooming and dressing oneself, preparing a meal, and performing basic household chores like doing dishes.

A person is diagnosed with MCI or *mild neurocognitive disorder* if their cognitive symptoms are relatively modest and do not interfere with their ability to perform activities of daily life independently, although it might take them longer, they might need to put more thought or effort into it (e.g., double-checking that they made the correct payments), and they might need some supports (like conscientiously keeping a calendar).

A person is diagnosed with *dementia* or *major neurocognitive disorder* if their cognitive impairments are more profound and severe enough that they are unable to function independently, even with supports like reminders and alarms. In mild forms of a dementia or major neurocognitive disorder, the person might be unable to perform IADLs, so somebody has to take over managing their finances, appointments, and prescriptions, but they can still perform their basic ADLs, like showering and dressing themselves. In moderate stages, the person cannot independently complete basic ADLs either, and they might need someone to prepare their meals and prompt them for hygiene. In severe stages, the person needs full care for their safety and comfort, as they might have difficulty walking, talking, or swallowing.

The second component of the diagnosis is the specific type of MCI or dementia the person is diagnosed with. This is often determined by the underlying neuropathology causing the cognitive decline. For example, MCI or dementia can be caused by Alzheimer's disease, cerebrovascular disease (like strokes), Lewy bodies disease, Parkinson's disease, frontotemporal lobal degeneration, a serious brain injury, a brain infection, and many other conditions. Other neurocognitive disorders are defined by the specific profile of symptoms—for example, primary progressive aphasias are diagnosed based on the presence of different patterns of prominent language impairments. Knowing the specific neurocognitive disorder a person is diagnosed with is important because the expected course of the condition and the available treatments are often different based on the specific type of disorder or their cause.

As I also mentioned in Chapter 3, there is variation in how professionals use these terms in practice. While *dementia* and *major neurocognitive disorder* both refer to cognitive impairments severe enough to interfere with independent functioning, historically *dementia* has been used when the cause is a neurodegenerative (progressive) condition; for example, we speak of dementia due to Alzheimer's disease, frontotemporal dementia, or dementia with Lewy bodies, all conditions that progressively worsen. For the most part—again, there is no absolute agreement—the term *major neurocognitive disorder* is broader, and can refer to conditions that are neurodegenerative but also those that might remain stable or potentially improve. For example, a patient who experiences a large stroke might be unable to perform activities independently for several months after the injury, meeting criteria for a diagnosis of a major neurocognitive disorder. However, with time and rehabilitation, their functioning might improve into a mild neurocognitive disorder range, because they are able to complete activities independently with some supports.

Similarly, the term *mild cognitive impairment* is used by some healthcare professionals specifically to refer to the mild cognitive symptoms caused by a neurodegenerative disease. Used this way, a diagnosis of MCI means that the person's symptoms will progress, eventually worsening into the dementia stage, when the person will not be able to function independently anymore. Because of this, you will often hear MCI defined as a "pre-dementia" stage, because the expectation is that the symptoms will worsen. The term *mild neurocognitive disorder* is typically used more broadly, regardless of whether the course is expected to be progressive or not.

It is important to address this because if you or a loved one is diagnosed with mild cognitive impairment, it will be critical for care planning purposes to clarify how the professional making the diagnosis is using the term—that is, whether they believe the condition is likely to worsen.

Important Questions to Ask

Here are some questions I suggest asking if you or a loved one are diagnosed with a neurocognitive disorder:

- What is the specific diagnosis? Are the symptoms at the MCI or dementia level? What disease or neuropathology is thought to be causing the condition?
- What is the expected course? Will the symptoms progress (worsen)? If so, what is the expected rate of progression? Is this a disease that typically advances slowly, or is rapid decline expected?
- How are symptoms expected to change? Are new symptoms (cognitive, emotional, behavioral) expected to appear?
- What treatment options, pharmacological and non-pharmacological, are available? What are the realistic expectations about treatment benefits? For example, will the treatments improve the symptoms, or just slow down their progression? What are potential risks and side effects? (Again, treatment options for many neurocognitive disorders are rapidly changing with the availability of new therapies and interventions.)
- Who will be part of the treatment/management team? Will you be referred to other specialists? How often will you have follow-up appointments?
- What counseling, educational, and other support services are available for the person with the diagnosis and their loved ones and care partners? For example, who can help with healthcare planning for the future?
- If you are interested, you can ask what clinical trials or other research opportunities are available. Participation in research can provide access to novel and experimental diagnostic tests and treatments.

As you process a new diagnosis and what it means for your life and your future, remember that millions of individuals are living with a neurocognitive disorder or helping to care for a loved one with a neurocognitive disorder: You are not alone. Because of the high prevalence of these conditions, improved understanding of their nature, and increased awareness of the challenges they present, resources and support are available to help you and your loved ones transition into life with a chronic brain disease.

* * *

Most of you reading this book are probably either not concerned enough about your cognitive lapses to seek evaluation, or might have already sought evaluation and were not diagnosed with a neurocognitive disorder. Not receiving a diagnosis is a relief but also somewhat frustrating, since you might have been left feeling your experiences have no explanation. That is the reason I wrote this book, to provide some possible answers and some possible solutions.

In modern healthcare, perhaps the most common frustration care providers and patients share is the lack of time to have in-depth, meaningful, and satisfying conversations about our health and healthcare. No patient should leave their appointment feeling they do not have enough knowledge or resources to take the most beneficial next steps. What I have attempted to do in this book is compile the information and advice that I wish I could give every patient, and hopefully provide something of a roadmap for you to work through as you attempt to build a healthier, happier, brain-friendlier life. I hope you have found it helpful.

Index

Printed in the United States
by Baker & Taylor Publisher Services